Lecture Notes in Earth Sciences

Editors:
S. Bhattacharji, Brooklyn
G. M. Friedman, Brooklyn and Troy
H. J. Neugebauer, Bonn
A. Seilacher, Tuebingen and Yale

Springer
*Berlin
Heidelberg
New York
Barcelona
Hong Kong
London
Milan
Paris
Singapore
Tokyo*

Stefan Hergarten
Horst J. Neugebauer (Eds.)

Process Modelling
and Landform Evolution

With 133 Figures and 14 Tables

 Springer

Volume Editors

Dr. Stefan Hergarten
Prof. Dr. Horst J. Neugebauer
Geodynamics – Physics of the Lithosphere, University of Bonn
Nussallee 8, D-53115 Bonn, Germany

GB
400.42
.M33
P76
1999

"For all Lecture Notes in Earth Sciences published till now please see final pages of the book"

Cataloging-in-Publication data applied for

Die Deutsche Bibliothek - CIP-Einheitsaufnahme

Process modelling and landform evolution : with 14 tables / Stefan Hergarten ; Horst J. Neugebauer (ed.). - Berlin ; Heidelberg ; New York ; Barcelona ; Budapest ; Hong Kong ; London ; Milan ; Paris ; Singapore ; Tokyo : Springer, 1999
  (Lecture notes in earth sciences ; 78)
  ISBN 3-540-64932-8

ISSN 0930-0317
ISBN 3-540-64932-8 Springer-Verlag Berlin Heidelberg New York

Typesetting: Camera ready by author
SPIN: 10691617   32/3142-543210 - Printed on acid-free paper

Preface

Landform evolution is driven by the interference of a broad spectrum of processes involving atmosphere, hydrosphere, biosphere and tectonosphere. Although we have direct access to this 'moving boundary problem', short term phenomena are rather complex, while the long term evolutionary aspects are often hidden by a lack of adequate observation. At present, much of understanding is conceptual rather than specific. Resolving these uncertainties is critical to understanding the nature of coupled physical processes operating on the evolution of landforms.

A recent workshop on **Process Modelling and Landform Evolution**, held at Bonn, Germany, in 1997 brought an interdisciplinary group of scientists together. Beside contributions from geography, especially from geomorphology and hydrology, there were contributions from applied mathematics, computer sciences, geology, geophysics, photogrammetry, and soil sciences. This wide range of disciplines was surprising for most of the participants, but it resulted in a lively discussion about the different methods and approaches.

According to the structure of the workshop, this book is organized in three parts. The first part deals with **Terrain Representation and Analysis**, which inevitalby links field studies and process modelling. Beside a general discussion of the apparent contradictions between the results from field studies and process modelling, various methods are discussed with respect to landform processes. Correlations between geometric properties of the relief indicate that processes or at least their parameters vary significantly on the regional scale. Techniques of automatic landform classification yield parameter sets for lumped models like large scale hydrological models. In contrast, physically based process models require an efficient adaptive representation of the relief in order to obtain an adequate resolution.

The second part is entitled **'Short Term Modelling'**; it is focused on the interactions of water with the relief. Approaches to surface runoff and small scale erosion processes are presented as well as approaches to subsurface hydrology (infiltration, lateral soil water flow) and its implications on slope stability. All these processes are linked with the relief in a twofold way; they are significantly affected by the morphology of the land surface, but also influence landform evolution over long times in a cumulative way.

Finally, the third part presents approaches to **Modelling Landform Evolution** itself. Obviously, the most striking problem is linking the wide range of scales contributing to landform evolution. Erosion processes enclose rill formation on the centimeter scale, hillslope processes of several meters, and fluvial processes of several kilometers. Together with landsliding and large scale tectonic processes they form the landscape. Large scale structures determine the boundary conditions of small scale processes, whereas the cumulative effect of small scale processes significantly contribute to landform evolution. Nesting of models on different scales is presented as one way of upscaling. As the occurence of landslides on different scales shows, small scale processes are not only able to form large structures in a cumulative way, but may also build up large events in a well–organized way. Concepts of self–organization are going to be established in landform evolution theory and apply well at least to landsliding processes.

Obviously, there is still a lot of open questions concerning landform evolution; a feasible universal model that captures all aspects of this complex phenomenon is not in sight. Hence, studies dealing with one aspect in detail and studies concerned with linking different scales will be of further interest; each can be improve our understanding of the complex processes and interactions governing landform evolution.

The workshop 'Process Modelling and Landform Evolution' was supported by the German Research Foundation (DFG) through the Collaborative Research Center 350 'Interactions between and Modelling of Continental Geo–Systems'.

The editors are indebted to I. Hattendorf and J. Schmidt for their help in preparing this book.

A board of 12 specialists served as anonymous reviewers of the contributing papers. Their help ist acknowledged with gratitude.

Bonn, July 1998

Stefan Hergarten
Horst J. Neugebauer

Table of Contents

List of Authors

M. G. Anderson, School of Geographical Sciences, University of Bristol, University Road, Bristol BS8 1SS, England, Fax: +44 117 928 7878

Th. W. J. van Asch, The Netherlands Centre for Geo–ecological Research, Dept. of Phys. Geography, Utrecht Univ., P. O. Box 80.115, 3508 TC Utrecht, The Netherlands, Fax: +31 30 254 0604, E–mail: T.vanAsch@frw.ruu.nl

L. P. H. van Beek, The Netherlands Centre for Geo–ecological Research, Dept. of Phys. Geography, Utrecht Univ., P. O. Box 81.115, 3508 TC Utrecht, The Netherlands, Fax: +31 30 254 0604, E–mail: R.vanBeek@frw.ruu.nl

C. Blendinger, Kiliansweg 3, 64823 Groß–Umstadt, Germany, E–mail: Christoph.Blendinger@tlc.de or blendi@netsurf.darmstadt.de

A. Braunmandl, Institute of Photogrammetry, University of Bonn, Nußallee 15, 53115 Bonn, Germany, Fax: +49 228 73 2712, E–mail: andre@ipb.uni–bonn.de

S. M. Brooks, School of Geographical Sciences, University of Bristol, University Road, Bristol BS8 1SS, England, Fax: +44 117 928 7878, E–mail: susan.brooks@bristol.ac.uk

N. J. Cox, Dept. of Geography, University of Durham, South Road, Durham DH1 3LE, England, Fax: +44 91 374 2456, E–mail: N.J.Cox@durham.ac.uk

M. Crozier, Institute of Geography, Victoria University of Wellington, P. O. Box 600, Wellington, New Zealand, Fax: +64 4 495 5127, E–mail: Michael.Crozier@vuw.ac.nz

R. Dikau, Institute of Geography, University of Bonn, Meckenheimer Allee 166, 53115 Bonn, Germany, Fax: +49 228 73 9099, E–mail: rdikau@slide.giub.uni–bonn.de

C. Droste, Institute of Photogrammetry, University of Bonn, Nußallee 15, 53115 Bonn, Germany, Fax: +49 228 73 2712, E–mail: christo@ipb.uni–bonn.de

T. Ennion, School of Geographical Sciences, University of Bristol, University Road, Bristol BS8 1SS, England, Fax: +44 117 928 7878

I. S. Evans, Dept. of Geography, University of Durham, South Road, Durham DH1 3LE, England, Fax: +44 91 374 2456, E–mail: I.S.Evans@durham.ac.uk

Th. Gerstner, Department of Applied Mathematics, University of Bonn, Wegelerstr. 6, 53115 Bonn, Germany, Fax: +49 228 73 7527, E–mail: gerstner@iam.uni–bonn.de

G. Gimel'farb, CITR, Department of Computer Science, Tamaki Campus, University of Auckland, Private Bag 92019, Auckland, New Zealand, Fax: +64 9 3082377, E–mail: g.gimelfarb@auckland.ac.nz

R. Hantke, Glärnischstr. 3, 8712 Stäfa, Switzerland

I. Hattendorf, Geodynamics – Physics of the Lithosphere, University of Bonn, Nußallee 8, 53115 Bonn, Germany, Fax: +49 228 73 2508, E–mail: hatten@geo.uni–bonn.de

K. Helming, Dept. of Soil Landscape Research, Center for Agricultural Landscape and Land Use Research (ZALF), Eberswalder Str. 84, 15374 Müncheberg, Germany, Fax: +49 33432 82280, E–mail: khelming@zalf.de

St. Hergarten, Geodynamics – Physics of the Lithosphere, University of Bonn, Nußallee 8, 53115 Bonn, Germany, Fax: +49 228 73 2508, E–mail: hergarten@geo.uni-bonn.de

M. J. Kirkby, School of Geography, University of Leeds, Leeds LS2 9JT, England

H. J. Neugebauer, Geodynamics – Physics of the Lithosphere, University of Bonn, Nußallee 8, 53115 Bonn, Germany, Fax: +49 228 73 2508, E–mail: neugb@geo.uni-bonn.de

G. Paul, Geodynamics – Physics of the Lithosphere, University of Bonn, Nußallee 8, 53115 Bonn, Germany, Fax: +49 228 73 2508, E–mail: paul@geo.uni-bonn.de

S. N. Prasad, Dept. of Civil Engineering, University of Mississippi, University, MS 38677, USA

N. Preston, Institute of Geography, University of Bonn, Meckenheimer Allee 166, 53115 Bonn, Germany, Fax: +49 +228 73 9099, E–mail: nick@slide.giub.uni–bonn.de

M. J. M. Römkens, USDA–ARS National Sedimentation Laboratory, P. O. Box 1157, Oxford, MS 38655, USA, Fax: +11 601 2322915, E–mail: poynor@sedlab.olemiss.edu

A. E. Scheidegger, Department of Geophysics, TU Wien, Gusshausstr. 27/128/2, 1040 Wien, Austria, Fax: +43 1 5044232, E–mail: ascheide@luna.tuwien.ac.at

J. Schmidt, Institute of Geography, University of Bonn, Meckenheimer Allee 166, 53115 Bonn, Germany, Fax: +49 +228 73 9099, E–mail: jochen@slide.giub.uni–bonn.de

H. Sommer, ESPE, ESPE–Platz, 82229 Seefeld, Germany, E–mail: sommerh@espe.de

P. Wilkinson, School of Geographical Sciences, University of Bristol, University Road, Bristol BS8 1SS, England, Fax: +44 117 928 7878

Part I

Terrain Representation and Analysis

The Need for Field Evidence in Modelling Landform Evolution

R. Dikau

Department of Geography, University of Bonn, Germany

Abstract. A recent debate within the international geomorphological community is related to a divergence between present process studies and landform evolution modelling. The paper argues that these models have to be based on some fundamental characteristics of landforms derived from field observations, including the spatial hierarchy of nested landform assemblages, the lifetime and temporal persistence of landforms, the 3–D sediment body of landforms, the 2–D surface of paleo landforms, and the spatial heterogenity of geomorphological process domains. It is suggested that different model types should be considered as requirements to modelling landforms and landform change in time including sediment storages, sediment budgets and event sequences.

Introduction

The workshop "Process Modelling and Landform Evolution" of the German Research Foundation (Deutsche Forschungsgemeinschaft) project "Interactions between and Modelling of Continental Geo–Systems" (SFB 350) addressed a problem which is related to a recent debate within the international geomorphological community about the fragmentation of the discipline between research modelling dynamic processes and long–term landform development. This paper is intended to focus some aspects of this debate in terms of discussions within the SFB 350 project concerned with geomorphological processes and landform evolution modelling.

The analysis of Earth surface systems in terms of their generic nature, genetic evolution, nature of geomorphologically relevant material and present processes is the main focus of geomorphological research. Within these objectives the analysis of the geometry and structure of landform surfaces, that is landform geomorphometry (or simply morphometry) and the stratigraphic record, play key roles in distinguishing the results of processes which have been operating over different scales in space and time, and thus identifying processes acting upon that landform as process background condition. This elucidates the dualistic character of the present landform (Chorley et al. 1984). Because landforms exist at different spatial and temporal scales they therefore carry two geomorphological meanings (Whiting and Furbish 1995):

1. Landforms are a result and a boundary condition of present formative processes acting on materials with specific properties. This means that we *predict*

a landform in a forward strategy from a set of geomorphological processes which can be modelled by physically–based approaches if appropriate data are available.

2. The inverse strategy is based on approaches *inferring* a set of formative processes from the appearance of a recent or a paleo landform and the material it is formed on. They are caused by paleo processes, which create paleo landforms, e.g. a terrace created by a lateral slope undercutting during the Pleistocene. To understand these landforms we need geomorphogenetical approaches modelling landform evolution combined with knowledge about background conditions e.g. the human impact, climate or tectonics.

In a holistic geomorphology both directions of thinking have to be applied to understand landform development. The specific problem of the inferential approach is, that often the depositional sediment is the only document of the past process sequence. Therefore, it is asserted in this paper that only the integration of "physically based geomorphic models" (Barsch 1995) with qualitative geomorphogenetical approaches can provide an adequate framework to strengthen convergent directions of geomorphological research.

Closely linked with these general statements is a current debate in geomorphology concerning the divergence between present processes and landform development (Rhoads and Thorn 1996, Stoddart 1997). Douglas (1982) emphasized this in the expression of the unfulfilled promise of earth surface process studies as a key to landform evolution. Brunsden (1993) described the apparent antagonism between functional surveys of present day process–response systems and historical and evolutionary studies which reconstruct the sequence of events and landscape responses by which the current stage came into being. This discussion has been reviewed and summarized especially by Beven (1996) and Haff (1996). Beven (1996) argues that models predicting landform evolution (e.g. Willgoose 1994) have significant problems in reproducing *particular* landforms with their individual histories and persistence of effects from past events and tectonic and/or climatic regimes.

The role played by modelling of geomorphological systems of different spatial and temporal scales is basically linking of the nested hierarchies of landform charactersitics in space with nested hierarchies of landforms in time (De Boer 1991). This can lead to general statements about the positive correlation between the size and the duration of existence of landforms (Ahnert 1988, Brunsden 1996). Kirkby (1996) emphasized in this context the role for theoretical models in geomorphology as qualitative thought experiments within a nested model hierarchy, where the temporal data come from stratigraphic sequences and the scaling problem is based on up–scaling of models from local to regional and global scales.

Spatial and Temporal Hierarchies of Landforms

Landform analysis has to deliver adequate descriptors of the generic character of the landform itself. We can apply a range of methodologies to our investigation, including

1. time represented by the stratigraphic record of the underground material,
2. the spatial distribution of landforms,
3. the substitution of a landform developing process sequence by a spatial association of recent landforms or landform components, which we call the ergodic principle or
4. the principle of actualism which is based on the transformation of processes acting under recent climatic conditions into a past environment for which the same climate has been reconstructed, or is inferred.

Landforms constituting the Earth surface are scale–dependent. This scale–dependency implies that landforms are ordered in spatial hierarchies of nested landform assemblages (Ahnert 1988, Dikau 1989, Pike and Dikau 1995, Brunsden 1996). All concepts of landform analysis are aimed at subdividing the spatial continuum into units which are related to the specific entity under investigation, e.g. a tectonic process or a sediment body. Based on this statement Brunsden (1993) suggested a possible solution to this problem by examining the lifetime and temporal persistence of landforms, which is "the way in which landforms and their process–response systems may survive for varying lengths of time to form a palimsest of systems". The term landform palimsest was suggested by Chorley et al. (1984) and means that smaller landforms are superimposed on larger ones and that older landforms are partially erased by younger landforms created by processes for which older landforms operate as a boundary condition.

One of the first approaches to the creation of a geomorphometrically–based landform classification model was developed by Albrecht Penck (Penck 1894, 1896) (see Beckinsale and Chorley 1991). Pencks model is based on the assumption that all landforms can be allocated to a set of some fundamental landform types including five hierarchical landform levels. Simple form elements build up basic landforms which can be aggregated to landform associations and finally to continental blocks. Fundamental surface forms may be classified into plains, escarpments, valleys, mountains, and cup–shaped hollows. Therefore, the fundamental unit is the form element which is the simplest unit in the model and which can be defined by curvature in plan and profile. Although Albrecht Penck did not introduce other morphometric parameters to define higher order units, his approach has influenced geomorphologists of this century, especially in Europe. Consistent landform classification models, which follow basically Albrecht Penck and which are based on quantitive morphometric parameters, were published by Jefremow (1949), Kugler (1974) and Speight (1988). These authors used, for the first time in landform analysis, purely morphometric parameters to classify landforms based on landform units, higher order units and parameters for quantification of landform patterns and structure.

The manner in which this linkage of landforms exists in landscapes was shown by Julius Büdel. One example of Büdel's conceptual model of geomorphological space (Büdel 1982: 34) is based on a nested three level hierarchy of an erosional process–response system of the Highland of Semien in North Ethiopia. The highland itself is a highly resistant tertiary landform which is controlled by its own system of shallow swales. The highland is dissected by deep canyons. The slopes of the canyons carry young V–shaped ravines encroaching from below. The ravines often do not meet the swales of the highland at the active rims and do not constitute any extension of them. Büdel argued that the two valley generations of swales and canyons are uncoupled and operate completely independently of each other constituting two different stages of temporal landform development. Space including landform size is linked with time, because smaller and, therefore, younger landforms (ravines and canyons) are superimposed on larger landforms (canyons and highlands). Furthermore, Büdel asserted independent process–form feedbacks in relation to landform size and age. Büdel's example is based on the hypothesis that geomorphological landforms obviously exist in a nested hierarchy of differentially adjusted landforms which reflect a different sensitivity to geomorphological change within nested negative feedback loops.

Despite of recent advances in computer technology there remains the fundamental question posed by Richard Chorley in 1972 of whether "the body of spatial techniques" enables "the building of dynamic spatial models of landforms which will undoubtedly form the keystone of much geomorphological work during the next decade" (Chorley 1972). There is a very clear message in this book underlying its aim of focussing attention on the body of spatial techniques: that to understand space is one of the prerequisites to understanding landform change in time. Systematic classification of landforms, their components and associations, and regional landform inventories are prerequisite to a deeper understanding of geomorphological form process–response systems (IAG 1995), and it is still an open question as to how to develop quantitative criteria for local, regional and global landform scales (Dikau 1990). A further aspect emphasized by Ritter (1988) is that in close connection with time (landform lifetimes and persistence, thresholds, frequencies, magnitudes, stratigraphic records etc.), "responses in surficial geomorphological processes are probably strongly controlled by spatial characteristics existing within the geomorphological setting" at different geomorphological scales, which also means "that structurally controlled units of the Earths surface have preferred rather than random dimensions". It is asserted that polyhierarchical approaches to geomorphometry, which are based on nested landform hierarchies as intended by Büdel (1982) or Chorley et al. (1984), could deliver a conceptual framework for classification and inventory of geomorphological systems in space and time (Dikau 1990).

A paleo landform is a geomorphological unit which was created in former periods of landform development. These landforms have been either eroded or covered by sediment depositions. Geomorphological research concerned with landform development uses stratigraphic records to reconstruct paleo process–

response systems and their boundary conditions. Because of the very limited availability of sediment–covered paleo–surface data, recent research has mainly focused on profile analysis. One example of this kind of landform analysis is described by Semmel (1989) who showed this situation in a profile through a paleo–hollow developed in loess near Frankfurt, Germany. A sequence of dated Tertiary and Pleistocene gravels, Pleistocene loess, different paleosoils and the Holocene colluvium shows that the evolution of this hollow happened in pulses consisting of different periods of stability and acitivity. Stability is characterized by soil development, while periods of activity show both the filling and erosion of the hollow. Loess was transported by aeolian processes during the glacial periods, and colluvial material has been transported by wash processes and accumulated during the Holocene, covering the older Pleistocene land surface. Different sequences are characterized by 3–D sediment bodies and several 2–D surfaces constituting the boundary layers of the sediment bodies. An objective which has to be included in geomorphological field measurements is therefore the quantification of the morphometry of covered landforms. These landform surfaces acted as a boundary condition in a paleo process–response system in combination with the 3–D sediment bodies which both constitute the basic components of the past landform.

Field evidence indicates that geomorphological processes do not operate homogeneously in space. During the Holocene period in some Alpine valleys high–magnitude catastrophic rock avalanches with more than 500 Mill. m^3 volume occurred on only 2 % of all slopes, in a recurrence interval of one or two events in 10,000 years. Obviously, some geomorphological processes are highly discontinuous in space if considered over specific time scales. In relation to space we are faced here with process domains, which are regions characterized by a given process and landform assemblage or a specific form–process relation or by a location of a series of process parameters, for example the morphometry of the slope–floodplain coupling (Thornes 1987) or rock strength properties . This implies spatially distributed areas of stability and activity which respond differently to an impulse of disruption (Rohdenburg 1970). For instance, we know in cuesta scarp locations in Germany, at the same spatial scale, highly active landslide slopes in close proximity to slopes which have been stable since the Younger Dryas. This spatial inhomogeneity of process activity has particularly been stressed by Brunsden (1996) who emphasized event sequences as the driving phenomenon of landform development. Event geomorphology, therefore, means to understand the frequency and magnitude of formative sediment movements in relation to both specific, climatic or tectonic boundary conditions and the specific scale under consideration (Crozier 1996, Hovius et al. 1997).

The magnitude–frequency concept has been well established in geomorphology since the publication of Wolman and Miller (1960) over 30 years ago. Many empirical studies have shown magnitude–frequency relationships which can be well described by a power law within a specific range of values. This was discussed in a summary of several empirical studies by Whalley (1984) for rockfalls and other landslide types. However, as Schumm (1991) emphasized, this concept is also an example of "how the relative amount of work done during different events is not necessarily synonymous with the relative importance of these events in forming a landscape" because although the high magnitude events carry very high sediment loads, they are infrequent, "and over a long period of time the most sediment movement will be accomplished during events of moderate magnitude and frequency".

Problems and Requirements of Modelling Landform Evolution

The previous section has attempted to address some aspects of geomorphological phenomena related to and derived from field evidences. Based on this discussion the following section reflects some problems resulting from these approaches in terms of process modelling and landform evolution and discusses some requirements for modelling landforms and landform change in geomorphology.

1. The analysis of nested hierarchies of landforms in space and time necessitates, as Kirkby (1996) suggested, a framework of a hierarchy of nested models ranging from site specific steady state assumptions to mass balance equations for cyclic time scales. This model framework would have to consider a problem included in theories of nested landform hierarchies in relation to specific temporal and spatial scales. The question is how processes acting on lower levels in space and time are coupled with higher level features. Richards et al. (1995) argued in response to the specific time and space scale hypothesis of Schumm and Lichty (1965) that time scales are clearly not discrete entities and that there must be feedback mechanisms between the processes that occur over different scales. In the case of the landform generations described by Büdel the open question is therefore how the uncoupled valley generations are connected in a process–response system or in other words, how and in which time scale the lower level processes disturb landforms of a higher order generation.

2. A further problem is related to the magnitude/frequency concept, which has been challenged during the International Association of Geomorphologists Conference in 1997 (Crozier 1997). In relation to a stronger focus on investigations of the complexity of landscape systems, e.g. in sediment budget approaches, it is obvious that research activities require a stronger emphasis on process linkages between components of geomorphological systems, where coupled processes fill and remove sediment storages at different magnitude–frequency relationships, which do not necessarily follow the power laws found in single process studies. As Crozier (1997) summarized "the challange now is to isolate and predict the landform product of intersecting process regimes, interrelated but all working to their own frequency and magnitude agendas". In close connection with the magnitude/frequency concept, in recent years a theory has been developed which shows interesting linkages to the landform evolution problem discussed. The concept of self–organized criticality (SOC)

(Bak 1996) is based on the hypothesis that complex behaviour in nature reflects the tendency of large systems with many components to evolve into a critical state, where minor disturbances may lead to events of all sizes. Most of the changes than take place through infrequent catastrophic events. Bak (1996) continues, that the evolution to this very delicate state would occur without design from any outside agent and that this critical state is self–organized, which means that the state is established because of the dynamical interactions among individual elements of the system. Following these arguments, the typical empirical magnitude/frequency relationship of geomorphological events (e.g. Whalley 1984) would therefore show a fractal behaviour or scale–invariance through a power law distribution. Hergarten and Neugebauer (1998) argued that this distribution suggests that SOC may play an important role in landform evolution, "at least if it is dominated by landsliding". Therefore the SOC concept may deliver a theoretical framework to integrate feedbacks between geomorphological processes acting in different scales. The application of the SOC theory in a landform evolution model by Hergarten and Neugebauer (1998) shows promising results and suggests that landslide processes may be considered as a SOC phenomenon.

3. A third problem is related to the recent debate about convergent and divergent directions of geomorphological research. Based on examples of divergent "bandwagons" in geomorphological research (functional, theoretical and applied) Slaymaker (1997) stresses the necessity for a unifying concept in geomorphology which would be "true to its intellectual roots". This concept is that of a sediment budget, which is defined as the quantitative description of the movement of sediment through a landscape unit. The sediment budget concept includes the understanding of sediment transport processes coupling different sediment storages. These storages build up the 3–D body of landforms with boundary layers to the atmosphere and to other sediment strorages (or landforms) created at the same or other time scales. The reconstruction of landform creation and removal necessitates therefore linking of nested models at different nested spatial and temporal scales. Besides the approaches related to spatial scales and upscaling procedures, the temporal aspect in modelling geomorphological systems has especially been emphasized by Slaymaker (1991). This concept also includes suggestions for the measurement of meso– and macro–scale geomorphological features.

4. As described above landform analysis has to recognize one of the main characteristics of geomorphological systems, that storages in the system may be filled and evacuated at different temporal scales. Therefore, landform modelling means not only the production of landform surfaces in real recent landscapes by deductive reasoning in an internally and numerically consistent way (Beven 1996). It means also modelling the development of individual landforms with particular persistences, which is the "length of time over which a landform survives as a diagnostic element of the landform assemblage" (Brunsden 1996), in relation to the climatic, tectonic and human setting. This, however, would require the understanding of the development of particular landforms in the past including the recognition that landform

creation may occur over a much shorter time span (formative event) than landform removal (relaxation time).

Based on these general statements and with reference to the categories given by Slaymaker (1991) it is suggested that the following models should be considered as requirements to modelling landforms and landform change in geomorphology:

1. Morphological models:
   - Quantification of static landform morphometry as result of long–term landform evolution on recent and paleo surfaces (morphometry of landforms of different scales and hierarchical levels, e.g. hillslopes, drainage basins, continental margins).
   - Models of static landform morphometry as background condition for recent geomorphological processes (e.g. slope angles, flow accumulation values, drainage networks).
2. Morphological evolutionary models:
   - Three dimensional models of the structure and volume of sediment storages of the landscape in relation to individual landforms.
   - Models to differentiate the storage age and the internal structure of the storage respectively using stratigraphic records in terms of magnitude/frequency and sequence of events creating and/or removing landforms.
   - Models to produce real landform morphometries through time by deductive reasoning of the dynamic change of landform surfaces and of sediment storages over time in a form–process feedback system.
3. Cascading models:
   - Models to connect the available sediment storages of landforms in the landscape in a sediment cascade approach based on sediment budget concepts.
4. Process–response models:
   - Physically–based process models. In terms of different time scales these models have to be ordered in a hierarchical structure since recent sediment transport models can be based on process studies whereas past transport models have to be based on boundary conditions with a much higher degree of uncertainty.

Acknowledgments. This paper is related to the projects B13 (landslides) and B15 (historical soil erosion) of the Collaborative Research Center (SFB) 350 "Interactions between and Modelling of Continental Geo–Systems" gratefully supported by the German Research Foundation (DFG).

References

Ahnert, F. (1988): Modelling landform change. In: Anderson, M.G. (ed.): Modelling Geomorphological Systems, 375–400, London.

Bak, P. (1996): How nature works: the science of self–organized criticality. New York.

Barsch, D. (1995): Gedanken über die Zukunft der Geomorphologie. Quaestiones Geographicae, 4:37–42

Beckinsale, R.P., and R.J. Chorley (1991): The History of Landforms Vol.3: Historical and Regional Geomorphology 1890–1950. London.

Beven, K. (1996): Equifinality and uncertainty in geomorphological modelling. In: Rhoads, B.L., and C.E. Thorn (eds.): The Scientific Nature of Geomorphology, 289–313, Chichester.

Brunsden, D. (1993): The persistence of landforms. Z. Geomorph., N.F., Suppl.–Bd. 93:13–28.

Brunsden, D. (1996): Geomorphological events and landform change. Z. Geomorph., N.F., 40:273–288.

Büdel, J. (1982): Climatic Geomorphology. Princeton.

Chorley, R.J. (1972): Spatial analysis in geomorphology. In: Chorley, R.J. (ed.): Spatial Analysis in Geomorphology, 3–16, London.

Chorley, R.J., S.A. Schumm, and D.E. Sugden (1984): Geomorphology. London.

Crozier, M. (1996): Magnitude/Frequency issues in landslide hazard assessment. Heidelberger Geogr. Arbeiten, 104:221–236.

Crozier, M. (1997): Concepts and issues in the study of the frequency and magnitude of geomorphic processes and landform behaviour. International Association of Geomorphologists Conference, 1997, Bologna (unpublished).

De Boer, D.H. (1991): Hierarchies and spatial scale in process geomorphology. A review. Geomorphology, 4(5):303–318.

Dikau, R. (1989): The application of a digital relief model to landform analysis in geomorphology. In: Raper, J. (ed.): Three Dimensional Application in Geographic Information Systems, 51–77, London.

Dikau, R. (1990): Geomorphic landform modelling based on Hierarchy Theorie. Proc. 4th Intern. Symposium on Spatial Data Handling, 23.–27. July, Zürich, 230–239.

Douglas, I. (1982): The unfulfilled promise: Earth surface processes as a key to landform evolution. Earth Surface Processes and Landforms, 7:101.

Haff, P.K. (1996): Limitations on predictive modeling in geomorphology. In: Rhoads, B.L., and C.E. THORN (eds.): The Scientific Nature of Geomorphology, 337–358, Chichester.

Hergarten, S., and H.J. Neugebauer (1998): Self–organized criticality in a landslide model. Geophys. Res. Letters, 25(6):801–804.

Hovius, N., C.P. Stark, and P.A. Allen (1997): Sediment flux from a mountain belt derived by landslide mapping. Geology, 25:231–234.

IAG (1995): International Association of Geomorphologists, Newsletter No. 12 (2/1995). Z. Geomorph. N. F., 39:265–268.

Jefremow, J.K. (1949): Versuch einer morphographischen Klassifikation der Elemente und einfachen Formen des Reliefs (russ.). Woprosy Geografii, Bd. 11, Moskau.

Kirkby, M.J. (1996): A role for theoretical models in geomorphology. In: Rhoads, B.L., and C.E. Thorn (eds.): The Scientific Nature of Geomorphology, 257–272, Chichester.

Kugler, H. (1974): Das Georelief und seine kartographische Modellierung. Diss. B, Martin–Luther–Universität Halle, Wittenberg.

Penck, A. (1894): Morphologie der Erdoberfläche. Stuttgart.

Penck, A. (1896): Die Geomorphologie als genetische Wissenschaft: eine Einleitung zur Diskussion über geomorphologische Nomenklatur. Comptes Rendas, 6. Int. Geogr. Kongress, London, Sektion C, 735–752.

Pike, R., and R. Dikau (1995, eds.): Advances in Geomorphometry. Z. Geomorph., N.F., Suppl.–Bd., 101, Stuttgart.

Rhoads, B.L., and C.E. Thorn (1996, eds.): The Scientific Nature of Geomorphology. Chichester.

Richards, K., N. Arnold, S. Lane, S. Chandra, A. El–Hames, and N. Mattikalli (1995): Numerical landscapes: static, kinematic and dynamic process–form relations. In: Pike, R., and R. Dikau (eds.): Advances in Geomorphometry, Z. Geomorph., N.F., Suppl.–Bd., 101:201–220, Stuttgart.

Ritter, D.F. (1988): Landscape analysis and the search for geomorphic unity. Bull. Geol. Soc. Am., 100:160–171.

Rohdenburg, H. (1970): Morphodynamische Aktivitäts– und Stabilitätszeiten statt Pluvial– und Interpluvialzeiten. Eiszeitalter und Gegenwart, 21:81–96.

Schumm, S.A. (1991): To Interpret the Earth – Ten ways to be wrong. Cambridge.

Schumm, S.A., and R.W. Lichty (1965): Time, space and causality in geomorphology. Am. J. Sci., 263:110–119.

Semmel, A. (1989): The importance of loess in the interpretation of geomorphological process and for dating in the Federal Republic of Germany. Catena, Suppl., 15:179–188.

Slaymaker; O. (1991): Mountain geomorphology: a theoretical framework for measurement programmes. Catena, 18:427–437.

Slaymaker, O. (1997): A pluralist, problem–focused geomorphology. In: Stoddart, D.R: (ed.): Process and Form in Geomorphology, 328–339, London.

Speight, J.G. (1988): Landform classification. In: Gunn, R.H., Beattie, J.A., R.E. Reid, and R.H.M. van de Graaff (eds.): Australian Soil and Land Survey Handbook, 38–59.

Stoddart, D.R: (1997, ed.): Process and Form in Geomorphology. London.

Thornes, J. (1987): Environmental systems – patterns, processes and evolution. In: Clark, M.J., K.J. Gregory, and A.M. Gurnell (eds.): Horizons in Physical Geography, 27–46, Totowa.

Whalley, W.B. (1984): Rockfalls. In: Brunsden, D., and D.B. Prior (eds.): Slope Instability, 217–256, Chichester.

Whiting, P.J., and D.J. Furbish (1995, eds.): Predicting Process from Form. Geomorphology, 13(3), Amsterdam.

Willgoose, G.R. (1994): A statistic for testing the elevation characteristics of landscape simulation models. J. Geophys. Res., 99(13):987–996.

Wolman, M.G., and J.P. Miller (1960): Magnitude and frequency of forces in geomorphic processes. J. Geol., 68:54–74.

Relations between Land Surface Properties: Altitude, Slope and Curvature

I. S. Evans and N. J. Cox

Department of Geography, University of Durham, UK

Abstract. The land surface is studied in terms of point properties, linear (flow–line and break–line) properties and areal properties. These show considerable structure and within–region predictability: the land surface is not (uni–)fractal but varies along multiple statistical dimensions. The simplest point properties are altitude itself and the first and second order derivatives, slope and curvature, of the altitude surface. Two components of each derivative are of proven value: it is best to separate the vertical (slope gradient and gradient change) from the horizontal (aspect and aspect change). The relationships between these five, some of which are not intuitive, are discussed here with illustrations from Germany.

Slope gradient usually varies with altitude, but not monotonically. The strength of the relationship varies, and in dissected plateaus the overall correlation is negative. The effects of mesoclimate (slope climate) mean that geomorphological processes vary with aspect, and there are a number of hypotheses of how this might affect gradient. Such signals are masked by considerable scatter on gradient–aspect plots, and it is important to use appropriate tools for relating linear to circular variables. Surface curvature (convexity taken as positive) is separated into vertical (profile, change of gradient) and horizontal (plan, change of aspect) components. The relationship between these is almost universally positive, but very weak: correlations are around $+0.2$, whether or not transformations to reduce the effects of extreme values are applied. As might be expected, profile convexity almost universally increases with altitude, although the correlation is rarely stronger than that of gradient with altitude. Plan convexity is least varied (closest to zero) on steep slopes; extreme but real values are found on floodplains.

1 Introduction

Modelling of the land surface should start with thorough study of the properties of actual land surfaces. These are in fact rather complex, and vary in many different ways (Evans 1984a, Evans and McClean 1995). Moore et al. (1991) and Schmidt and Dikau (1998) have listed numerous properties, definable at points on the surface, which are of considerable hydrological or geomorphological relevance. Here we consider the five most general properties of this sort, the derivatives of the surface (Evans 1972, 1980); we focus on relations between these derivatives. Examples are drawn primarily from a number of German DEMs made available by Jochen Schmidt (Department of Geography, University of Bonn): results from these are set in a wider context in Tables 3 to 5. Table 1 gives the attributes and locations of the German data sets analysed here. They were produced by stereophotogrammetry (Weiherbach) or by interpolation from digitised contours (Eyach, Ohebach, Wernersbach).

Table 1. Characteristics of the German DEMs illustrated (Note: each area is rectangular, going beyond the basin named).

Name	Vert. res. m	Grid mesh m	Points	Area points	km^2	Relief	Location
Eyach	1.0	50	167×167	27225	68.06	high	Nn. Schwarzwald
Weiherbach	0.1	12.5	329×177	57225	8.94	low	Bruchsal, Kraichgau
Wernersbach	0.1	10	409×281	113553	11.36	mod.	Near Dresden
Ohebach	0.1	10	161×161	25281	2.53	low	N. of Goslar

In addition to providing background on surface derivatives and demonstrating the complexity and variability of the land surface, in this article we aim to show that some relationships between surface derivatives are universal, while others are specific to certain types of topography, and vary between such types.

2 Derivatives

Fig. 1 shows the derivatives of the altitude surface which have been found most useful by geomorphologists and others. The first derivative is slope, a vector with two components, often expressed by two angles: gradient (slope angle) and aspect (azimuth). Slope geomorphologists conventionally express these in degrees, respectively from the horizontal and clockwise from north. Engineers commonly express gradient as the tangent. The sine is relevant to assessment of the effects of gravity.

The second derivative, surface curvature, is a symmetrical 2×2–tensor and thus consists of three independent components: so far geomorphologists and hydrologists have found use only for the two components shown in Fig. 1. Compared with slope, there are more variations in definition or expression of surface curvature. We find it useful to separate the vertical (down a slope line) from the horizontal (along a contour) curvature, using the convention that convexity to the atmosphere is positive and concavity is negative. Hence 'profile convexity' and 'plan convexity' seem appropriate terms, and both may be expressed as degrees per 100 m (Young 1972). In geomorphological mapping, German authors have preferred to express these variables as radii of curvature (Barsch and Liedke 1980) with 6, 300 and 600 m taken as critical radii of curvature for crest and slope segments and hillocks. These correspond to 955, 19.1 and 9.5° per 100 m. Young (1972: 176) used radii of 50 and 500 m, corresponding to 116 and 11.6°

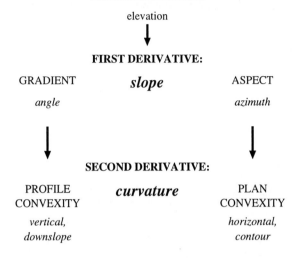

DERIVATIVES:

ALTITUDE SURFACE

elevation

FIRST DERIVATIVE:

GRADIENT	*slope*	ASPECT
angle		*azimuth*

SECOND DERIVATIVE:

PROFILE CONVEXITY	*curvature*	PLAN CONVEXITY
vertical, downslope		*horizontal, contour*

Fig. 1. Simple derivatives of the altitude surface. Alternative terms are given in italics.

per 100 m, to separate 'almost straight' slopes from 'slightly' or 'notably' convex (or concave) in plan. Either way, several classes are provided for mapping, but a major drawback for statistical analysis is that the radius is indeterminate for straight slopes which (as shown below) are the most common. Moreover, for very low curvatures radius is unduly sensitive to small errors, meaning that similar slopes may have very different values of this measure of slope curvature.

The third component of the second derivative is the mixed derivative: it relates to the local twisting of the surface and might be expressed by aspect change downslope or by the angle between the line directly downslope (slope line, line of greatest slope) and the line of maximum curvature. The effects of such a component could be a subject for future research, but it may well be that this concept is too abstract to be of much importance for surface processes. Other definitions include transverse curvature (Zevenbergen and Thorne 1987), but we prefer the separation of profile and plan, giving clear relations to the first derivative; profile convexity is the rate of change of gradient, and plan convexity is the rate of change of aspect. Eyton (1991) calculated related derivatives from finite differences. Shary (1995) considered relations between various types of geometric curvature in more detail. Evans (1972: 67) defined 'local height' and 'local convexity' as combined measures of convexity: recently Blaszczynski (1997) has found similar measures useful in subdividing topography at various scales.

The 'five derivatives' (including altitude itself, the 'zeroth derivative') were clearly recognised in the context of soils by Aandahl (1948), and taken up in geomorphology by Curtis, Doornkamp and Gregory (1965) and Speight (1968).

Fully quantitative definitions were given by Krcho (1973), and Evans (1972, 1979) computerised the calculations. Gradient is clearly of greatest importance for many applications (Demek and Embleton 1978, Crofts 1981). Altitude and aspect affect climate or slope climate, with major impacts on solar radiation receipt, temperature and water availability. The effect of aspect increases with gradient.

Infiltration is affected by profile and plan convexity as well as by gradient and climate. Profile convexity affects acceleration of surface flow, and plan convexity affects convergence and divergence and thus depth of flow, leading on to considerations of slope position and area drained (Schmidt et al. 1998). Slope and curvature as defined here have been applied to slope instability by Lanyon and Hall (1983) and to fault recognition by Florinsky (1996). Pennock, Zebarth and de Jong (1987) applied them to soil distribution and showed that the wetness of sites concave in profile or plan gives deeper A horizons and greater depth to calcium carbonate accumulation: in their Saskatchewan study area, gleyed horizons were found only on sites concave in both profile and plan.

In principle, all five variables are attributes of points, and can be estimated for any point on the land surface for which a tangent plane can be defined. In practice, whether measured in the field or from a DEM, they are measured over a small neighbourhood, which overcomes arguments as to whether the land surface is truly differentiable, i.e. smooth (Florinsky 1996: 106). With a gridded DEM, slope can be calculated by relating a point to one neighbour in each orthogonal direction. However, this treats the data as being error–free, and slope values relate to horizontal positions half–way between the altitude values: curvature values apply to a third set of positions. This in turn hinders the interrelation of altitude, slope and curvature. Instead, we have adopted a local fitting scheme, with slight smoothing, which provides slope and curvature for the same locations as the initial altitude values, permitting the five derivatives to be interrelated.

Fig. 2 is a schematic representation of a local quadratic fitted to altitude values in a 3×3 neighbourhood of a square grid. The quadratic has six coefficients, calculated from polynomial expressions which are simplified by the gridded nature of the data. Because there are nine altitude values, the surface does not exactly pass through the data points. Three are spare and the fitted surface is a smoothed version of the original. This is desirable because no data are error–free. Even if the altitudes were exact, for the precise locations specified, they are not recorded to many decimal places. It is common to store values to the nearest metre, or tenth–metre: exactly–fitting functions will therefore be distorted, especially in areas of low relief. Thus the degree of smoothing applied here is well justified: indeed in some circumstances the smoothing may not be sufficient to suppress artefacts.

Guth (1995) discussed the characteristics of five different slope calculation methods. Gradients he calculated by the 'eight neighbours unweighted' method, which corresponds to that used here, average some 78 % of those for 'steepest adjacent neighbour': this is similar to the effect of doubling grid mesh. Steepest adjacent neighbour gives the greatest slope in the vicinity and is thus positively

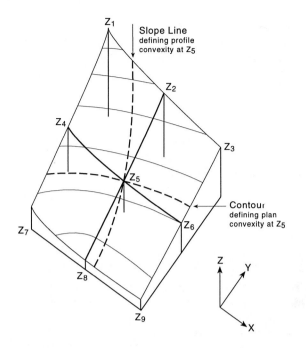

Fig. 2. The local quadratic fitted to a 3×3 neighbourhood of gridded altitudes.

biased in terms of gradient at the initial point; it permits only eight aspects and it does not relate to curvature at the same point. Correlations between gradients calculated from eight neighbours unweighted, eight weighted, and 'four closest neighbours' (the most commonly used algorithm) exceed +0.99. Carter (1992) showed that eight–neighbour methods gave lower gradient RMSE (root mean square error) and better aspect distributions than four–neighbour methods. Skidmore (1989) has shown the superiority of the local quadratic over various alternative algorithms; Guth (1995) showed that it produces better histograms of gradient and aspect. In any case, an exactly–fitting partial cubic as used by Zevenbergen and Thorne (1987) provides the same estimates of the five derivatives at the central point as does the quadratic; there are currently no uses for the cubic terms. We argue that the quadratic provides a better definition of slope and curvature than any alternative, and it permits derivatives to be interrelated at given locations, at a specific scale.

Table 2 gives some summary statistics for the frequency distributions of four derivatives (aspect, a circular variable, is treated separately) over the four German sample areas. These vary from the moderately high relief (altitude standard deviation 120 m) of the Eyach area to the much flatter Ohebach area. Both Ohebach and Wernersbach areas consist largely of low terraces.

Table 2. Statistics for surface derivatives of four German DEMs: units are given wherever appropriate.

Name	Altitude	Gradient	Profile Convexity	Plan Convexity
Mean:	m	°	°/100 m	°/100 m
Eyach	762	12.1	0.22	0.33
Weiherbach	196	5.2	−0.01	−0.58
Wernersbach	387	3.7	0.05	−0.01
Ohebach	142	2.9	−0.23	−3.30
Standard Deviation:	m	°	°/100 m	°/100 m
Eyach	120.5	7.3	7.44	87.2
Weiherbach	24.5	3.5	11.59	310.3
Wernersbach	27.1	2.9	9.06	150.6
Ohebach	9.6	2.4	10.65	257.9
Skewness:	dimensionless			
Eyach	−0.52	0.24	−1.05	−1.27
Weiherbach	−0.43	1.67	−0.36	−17.56
Wernersbach	−0.75	2.30	−1.21	−3.67
Ohebach	−0.90	2.59	−1.55	−10.80

Number of zero gradient points:					
	Summit	Plain	Saddle	Valley	Pit
Eyach	0	3	2	2	3
Weiherbach	5	3	2	2	3
Wernersbach	0	130	1	0	2
Ohebach	0	9	1	0	0

3   Complexity, Scale and Extent

Before considering relationships between the five derivatives, a number of important concepts must be mentioned. First, the land surface is not as random, or as fractal, as many expect; we shall return to this near the end. A long history of geomorphological writing has demonstrated the presence of many structures and patterns. Few of these are even approximately regular; the surface varies in many complex ways and it should be analysed with an open mind. As Neugebauer stated in his introduction to the meeting, the complexity of the land surface is such that we should strive for understanding before prediction. Some models can be truly Procrustean beds into which an ill–fitting reality is squeezed or stretched.

Second, the scale of the grid mesh for which data are available affects many results, but not all. In his talk, Dikau gave 50 to 200 m as the main (linear) scale of interest to geomorphologists. This 'mesorelief' may be defined by altitude grids between 10 and 100 m in mesh. Nogami (1995) suggested that grids of 50 m and finer are required for the analysis of Japanese topography. Zhang and Montgomery (1994) considered DEMs with meshes of 2 m to 90 m, for two small well–dissected catchments in California and Oregon: they suggested that a 10 m grid provided improved hydrological and geomorphological modelling, and finer grids gave only marginal improvement. The appropriate mesh is "somewhat finer than the hillslope scale identifiable in the field": this is "to adequately simulate processes controlled by land form" (Zhang and Montgomery 1994: 1026–27). 50 m is the basic grid mesh chosen for DEMs of France, 30 m for the U.S.A. Meshes much coarser than 100 m are of little value for analysis of mesorelief and hydrological pathways, but are of value in tectonic studies (e.g. Fielding et al. 1994).

Variations in grid mesh have no effect on the altitude distribution, but are important for first and second derivatives. As finer meshes bring in more details, the mean and standard deviation of gradient are increased (Evans 1979, Chang and Tsai 1991). The gentlest calculable slope is an inverse linear function of grid mesh. The shape of the gradient distribution (skewness and kurtosis) is only moderately affected. Profile and plan convexity include more extreme values; their standard deviations increase greatly from 100 m mesh to 10 m mesh data, although their mean values should be unaffected. Hence comparisons should be made only between data sets with the same grid mesh. Only altitude and aspect statistics can be safely compared across grid meshes. Even at the same grid mesh, results may differ for data sets derived by different techniques: this needs further investigation. The techniques and definitions used are scale–free in the sense that they can be applied to surfaces at any scale, from microrelief and engineered surfaces to global relief, but the meaning and the relative importance of derivatives change with scale.

Third, the extent of area studied is important, because the land surface is a non–stationary local variable. Sometimes authors have embarked on spectral or fractal analysis of data sets which are too limited for reliable estimation of the features or parameters of interest. If the regional relief is of interest, the data need to extend at least over a few valleys and ridges or hills, or features will be omitted and no estimate of vertical or horizontal landscape scale can be made. Despite these problems, Evans (1998) shows that moment statistics, vector statistics and correlations are replicable in relation to resampling with displaced grids, even for small DEMs, with the exception of skewness and kurtosis of profile and plan convexity. Both the convexity measures are sensitive to data quality, and plan convexity often gives extreme values, some of which are real.

4   Data Accuracy

Attention must be paid to the origins and accuracy of the data used. Data surveyed on a grid in the field are expensive and rare, but do exist for some engineering projects and some microrelief studies. More commonly, altitude data are surveyed by photogrammetry: the most accurate results are obtained by taking equi–spaced points along parallel profiles (Petrie and Kennie 1990). In terms of the effort involved it makes sense to space points closer along scan lines than between, producing an oblong grid; a square grid is made later by interpolating additional profiles. Even so, direct gridding is probably more accurate than interpolating an entire grid from digitised photogrammetric contours which, despite technical advances in radar interferometry and laser profiling, remains the most common means of production of the DEMs currently available for large areas (Carrara et al. 1997).

It is always possible to interpolate extra data, but we must remember that the information content has not been increased. Thus some data sets do not truly have the high resolution (small grid mesh) claimed, and geomorphologists are in danger of analysing the results of interpolation rather than the actual land surface. Some gridded DEMs are produced through GIS, via an irregular triangle stage: these should be viewed with suspicion (Wise 1998). Carrara et al. (1997) demonstrate how the ArcTin routine (as applied to three Italian test areas) fills valleys and depressions, flattens summits and ridges, and favours the altitudes of contours rather than intermediate altitudes. Their own routine (MDIP) and the Intergraph MGE5 TIN generator avoided these problems, but all four gridding techniques tested produced artefacts on the broad valley floor of the Adige.

With modern techniques of visualisation gross and moderate errors are soon obvious, as spikes on perspective views or as grid–mesh–sized 'volcanoes' in contour plots. Smaller errors are more difficult to detect, but persist in partly–interpolated data. The effects of poor data quality include striping of gradient and curvature maps, and spikiness of altitude and aspect histograms. Brown and Bara (1994) showed that smoothing was desirable for USGS 30 m DEMs, before calculation of derivatives: because the data are interpolated from photogrammetric scan lines 90 m apart, gradient and curvature maps show banding, and apparent roughness is greater along N–S profiles. Unweighted smoothing by a 3×5 filter was sufficient to remove this effect. Satellite–derived DEMs show patchiness and need even more smoothing (Giles and Franklin 1996).

One especially common error is bias toward the altitudes of the contours digitised. This is very prominent if bilinear interpolation is used (Evans 1984b); more complex, multidirectional techniques are necessary and the fact that points are along contours rather than randomly distributed should be used explicitly in the interpolation. However sophisticated the program used, flat areas may be produced when only one contour is found within the search radius. Many users complacently accept a default radius, but as contour density varies very different search radii may be required in different parts of a region.

Figs. 3 and 4 show detailed histograms (spike plots; Cox and Brady 1997) of altitude and gradient for two of the areas, Eyach and Ohebach. The Eyach area has a broad mode at altitudes of 750 to 940 m, the upper surface of the Schwarzwald massif. At lower altitudes, which are mainly valley–sides, frequency declines steadily to the minimum of 418 m, giving an overall skew of –0.52. The gradient distribution has a broad spread from 1 to 22°, with a short tail to the steepest gradients.

Ohebach, a much smaller area, also has negatively skewed altitude frequencies, with a pronounced maximum from 140 to 155 m, a long tail toward lower altitudes, and a much shorter upper tail: its overall relief is much lower than Eyach. Gradients of the extensive terrace areas are below 5°, but a few slopes in the narrow valleys are between 15 and 30°, giving a strong positive skew overall. These histograms are the most detailed to be relatively smooth between classes: very detailed histograms show spikes (single classes with many more values than adjacent classes) due to discretisation. This is discussed further in relation to aspect distributions.

In Figs. 5 and 6, frequency distributions of aspect are shown in detail, in one–degree classes. When some plots showed spikes at the eight cardinal directions (and several others), further aspect plots were produced by different bands of gradient. It was thus established that the spikes are limited to slopes below a threshold gradient, which varies between data sets. Figs. 5 and 6 represent these thresholds by separating the aspect plots for gradients greater and less than the threshold value. The spikes are confined to gradients below 15° for the relatively coarse Nupur data set (not illustrated) and below 5° for Eyach (50 m mesh, 1 m vertical resolution). They are absent for some high–relief French 50 m grids recorded at 0.1 m vertical resolution, but not for the 10 m grids here, despite their high resolution: the threshold is 3° for Weiherbach and Ohebach, but 5° for Wernersbach (Fig. 6). Hence they are not simply a function of coarse vertical resolution; finer grid meshes need to be accompanied by finer vertical resolution, or the threshold will rise.

Carter (1992) came to a similar interpretation, calculating slope from finite differences (four neighbours): where height differs by 1 m between neighbours on a 30 m grid, only eight aspects are possible. Testing with mathematical surfaces, he showed that the ratio between vertical increment and grid mesh controls errors in slope (aspect and gradient): aspect errors can be tens of degrees. For a 2–degree slope, altitudes rounded to 1 m permit only 16 possible aspects: rounding to 0.1 m permits about 100: and 0.01 m permits all 360 one–degree

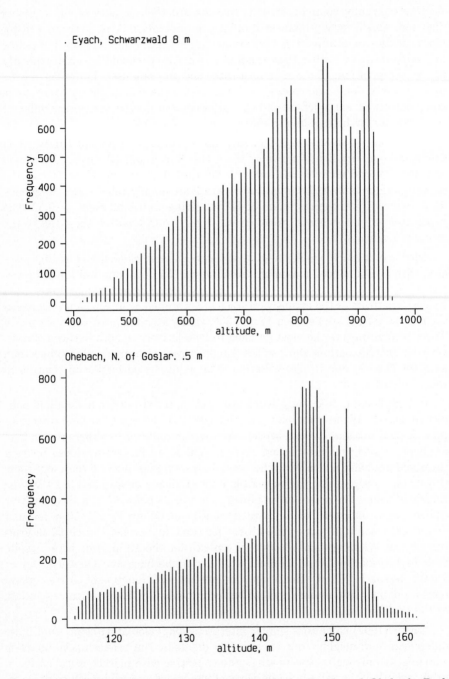

Fig. 3. Detailed histograms (spike plots) of altitude for Eyach and Ohebach. Both altitude frequency distributions are negatively skewed.

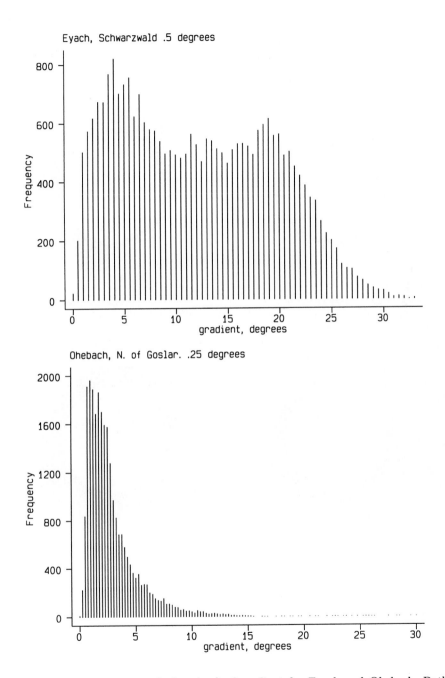

Fig. 4. Detailed histograms (spike plots) of gradient for Eyach and Ohebach. Both gradient distributions are positively skewed, but in addition Eyach shows a secondary mode at 19 degrees.

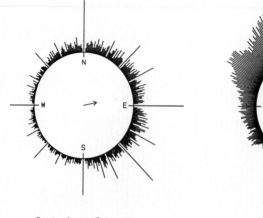

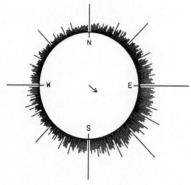

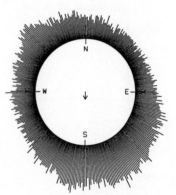

Eyach, slopes <5 degrees
mean direction 78.2 : vector strength 0.273

Eyach, slopes >5 degrees
mean direction 20.7 : vector strength 0.200

Weiherbach, slopes <3 degrees
mean direction 127.2 : vector strength 0.207

Weiherbach, slopes >3 degrees
mean direction 181.0 : vector strength 0.143

Fig. 5. Detailed aspect histograms (1 degree classes) for (left) low and (right) high gradients. The numbers of values included are 5945 and 21270 (Eyach); 15762 and 41448 (Weiherbach). Frequency distributions are fairly smooth for steeper slopes, above 3 or 5 degrees for these data sets. For the gentler slopes, spikes indicate a bias toward the eight major compass points and at least eight other aspects. The central arrow indicates vector mean aspect direction and strength (proportional to arrow length.

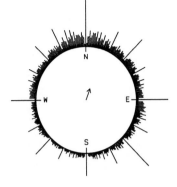

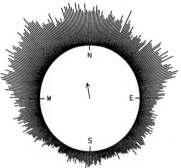

Wernersbach, slopes <5 degrees
mean direction 20.0 : vector strength 0.222

Wernersbach, slopes >5 degrees
mean direction 349.7 : vector strength 0.288

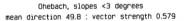

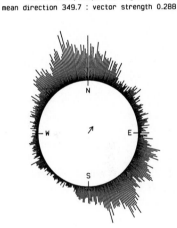

Ohebach, slopes <3 degrees
mean direction 49.8 : vector strength 0.579

Ohebach, slopes >3 degrees
mean direction 37.1 : vector strength 0.161

Fig. 6. Detailed aspect histograms (1 degree classes) for (left) low and (right) high gradients. The numbers of values included, are 88759 and 24661 (Wernersbach); 17088 and 8183 (Ohebach). Frequency distributions are fairly smooth for steeper slopes, above 3 or 5 degrees for these data sets. For the gentler slopes, spikes indicate a bias toward the eight major compass points and at least eight other aspects. The central arrow indicates vector mean aspect direction and strength (proportional to arrow length.

classes, although with four–fold variation in occupancy. Wise (1998) also finds aspect errors of tens of degrees, and considerable variation between algorithms: histograms based on local quadratics were less spiky than those based on Zevenbergen and Thorne's (1987) partial cubic. As a test of DEM accuracy, Wise recommends analysis of aspect rather than of altitude. Convexity can be even more demanding.

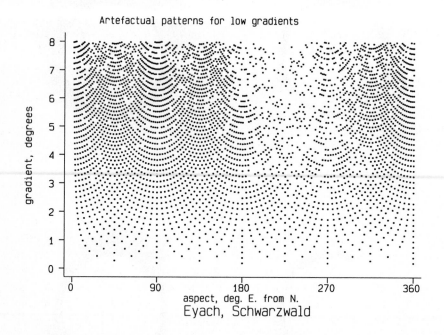

Fig. 7. Gradient versus aspect for Eyach. The festoon pattern reflects the limited number of combinations of aspect and gradient which are possible for grids with a 1 m altitude resolution and a 50 m horizontal mesh, especially on low gradients (below 5 degrees).

Where neighbouring grid points either have the same altitude or differ by one or two recording units, the number of aspect and gradient values possible is limited whatever algorithm is used for calculation. For local quadratics this is clearly portrayed in Fig. 7, a scatter plot of gradient against aspect for low–gradient slopes in the Eyach area. Vertical lines of points occur every 45°, with those every 90° being more populated. Unfortunately, the plot does not distinguish the number of values plotted at the same position. Most points for gradients below 5° fall into a festoon pattern, with a strong 90–degree and a weaker 45–degree periodicity. Evidently the number of possible gradient–aspect combinations declines drastically with decreasing gradient, thus producing the spikes. Going back to Figs. 5 and 6, the intermediate spikes are at 27, 45 and 63°

in the northeast quadrant, and at corresponding values in the other quadrants. In each area, 27 and 63° are favoured rather than 22.5 and 67.5°. This effect must arise because 27 and 63° are the arctangents of 0.5 and 2.0 respectively.

Figs. 5 and 6 show that the observations producing a spike have been displaced only a short distance; there is usually a deficit in the adjacent one–degree classes, and sometimes in classes two degrees from the spike. A small bias to the eight main directions is discernible also for plots with classes four, five or ten degrees wide, but not for fifteen degrees. Thus the error involved is tiny, certainly in a field where aspect is often taken to the nearest multiple of 45° (various comparisons, for cruder algorithms, show that only a minority of values are placed in the correct class). The error is also less serious than the altitude 'bias to contour' noted above. Nevertheless, this discretisation is an artefact, one that is attributed here to data resolution, and hence it provides a lesson worth noting. It is hypothesised that discretisation would hardly occur for data (of these grid meshes) recorded to vertical resolutions of 0.01 m or finer. Although it is not attributed to the technique used for calculating derivatives, discretisation might be reduced by augmenting that technique with an initial gentle smoothing and storing the results at 0.01 m resolution.

5 Gradient Relationships

From their definitions, the five derivatives might be expected to be relatively independent of each other: we do not expect any of them to be sufficiently well predicted from the others that it could be omitted. In random topography, correlations between them are expected to be low, for example the strongest altitude–gradient correlation for eleven fractal surfaces generated by McClean (1990, Table 7.2) was –0.13. In real topography, however, some weak general tendencies are observed, and moderately strong relationships are found for particular types of topography.

It is often observed that gradient increases with altitude; indeed altitude has been used as the relief factor in some models of sediment yield. The existence of the Tibetan Plateau should be a salutary warning here (Fielding et al. 1994): Summerfield (1991) has shown that a correlation of sediment yield with altitude should not be assumed and that altitude cannot be used as a surrogate for mean gradient or for relief (variation in altitude). The relationship of gradient to altitude is region–specific and correlation commonly varies between +0.5 and –0.5, but in general the relationship is curvilinear (parabolic). In lowland areas the highest gradients may be in isolated hills or in an upland fringe, giving strong positive relationships for example in the Wessex region of England (Table 3). Also in fluvially dissected, high–relief areas such as the Ferro basin in Calabria (Evans 1990, Fig. 2.5), in the Appalachian Ridge and Valley Province (McClean 1990: 166) and in areas with broad valleys, the concentration of low gradients at low altitudes gives a moderately positive relationship (around +0.5). Some areas have a negligible linear relationship, often because gradients are highest at an intermediate altitude. But areas of dissected plateaus or dissected terraces have negative relationships between gradient and altitude, and that is the case in the German data sets considered here.

Table 3. Ranges for correlations of gradient and profile convexity with altitude: the first two rows and the fourth are based on McClean (1990), p. 166, with some omissions and additions (see also Evans (1980), Table 2). Initial zeros are omitted for clarity. The Wessex results (t) are after root sine transformation of gradient and arctangent transformation of profile convexity.

		Correlation Ranges			
		Altitude – Gradient		Altitude – Profile Conv.	
DEM mesh:		min	max	min	max
9 @ 100 m	(various)	−.35	.55	.13	.47
10 @ 50 m	(France, England)	−.49	.50	.14	.36
54 @ 50 m	Wessex	−.06t	.65t	.09t	.28t
4 @ 30 m	(USA)	.17	.52	.06	.12
Eyach	50 m	−.45		.34	
Weiherbach	12.5 m	−.01		.21	
Wernersbach	10 m	−.54		.11	
Ohebach	10 m	−.35		.22	
Specific Landform types:		min	max	min	max
75 Cumbrian cirques @ 50 m, 5–95 % range only		.14	.82	.20	.71
7 Cumbrian drumlins @ 50 m		−.52	.42	.38	.75

In the Eyach area (Fig. 8), the linear correlation of −0.453 reflects a very broad spread, with the greatest density of points falling on low gradients at high altitude. Discretisation is not a problem for these variables, except at the lowest gradients. The steepest gradients are found over a broad range of middle altitudes (550 to 870 m), and gentle gradients are rare at 620 to 730 m. Hence a parabolic relationship (R = 0.536) is a clear improvement over a linear one: it predicts lowest gradients at highest altitudes and clearly should not be extrapolated beyond this altitude range. The coefficient R is the square root of the multiple correlation coefficient; as such it varies between 0 and 1. It is used here so that compound relationships can be compared with bivariate ones.

The plot for Ohebach (not illustrated) shows lower gradients at all altitudes. Gradients over 16° are found only from 120 to 140 m, out of a total range of 114 to 161 m: below 140 m few slopes have gradients less than 1 degree. Again, the relationship is parabolic (R = 0.39). In the Wernersbach area of Saxony there is a steady trend toward lower gradients at higher altitudes. The negative correlation is reinforced by the steepest slopes, including all those above 17°, being at lower altitudes. This gives an overall correlation of −0.544, but gradient variability is much greater at low altitudes.

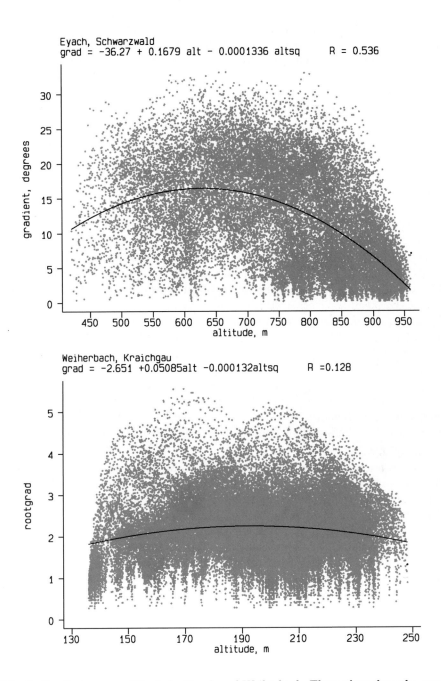

Fig. 8. Gradient versus altitude for Eyach and Weiherbach. These plots show the common parabolic relationship, with greatest gradients at intermediate altitudes.

Except for Eyach, gradients have strong positive skew (Table 2) and are best treated with a transformation such as square root. For Weiherbach (Fig. 8), the weak parabolic relationship is dominated by the scatter, but confirms the somewhat steeper gradients of mid–altitudes. Addition of the parabolic term (altitude squared) provides a modest but clear strengthening of correlation for the four German DEMs. All four areas have the characteristics of dissected plateaus or dissected terraces.

In general, the steepest slopes are rarely at the extremes of altitude: this might not be the case, however, if areas with coastal cliffs were included. Conversely, the gentlest slopes are most common at either extreme of altitude, depending on the topography. For fractal surfaces the altitude–gradient correlation is near–zero: in a study by McClean (1990:180), it varied between –0.13 and +0.11, with no relation to fractal dimension. Hence, while the general curvilinearity of scatter plots must not be forgotten, correlations stronger than 0.15 are likely to be real characteristics of the study areas.

Traditionally, altitude–gradient plots have been generalised by a clinographic curve, plotting the mean gradient for each altitude band (e.g. Hormann 1971:68–70). This loses too much information, but a middle course is possible by combining the mean with other statistics such as mean plus/minus one standard deviation, and maximum and minimum. Alternatively, the median and selected quantiles can be plotted. In Fig. 9 we have summarised the scatter by plotting five percentiles of gradient for a series of equal bins (classes) of altitude: the 10th, 25th, 50th, 75th and 90th percentiles, i.e. the median, the other quartiles and the outer deciles (Cox 1997). For Ohebach (as for Eyach), their trends are consistently parabolic, with maximum gradient percentiles around 125 m (but relatively high gradients in the highest altitude bin). For Weiherbach there is more variation, but maxima around 175 m and 225 m are stable across the percentiles.

The relation of gradient to aspect can also be treated in terms of averages for bands (Hormann 1971:73–87), or percentiles as in Fig. 9, but it is best summarised by a Fourier series (sine, cosine) regression of whatever order is required to capture the relevant detail. Most hypotheses on the effects of climate require a first–order regression (a single sine and a single cosine term), whereas those on the effects of geologic structure often require second or higher orders. Higher Fourier terms may either define subsidiary modes, or sharpen up or blunt the main modes or troughs. Wind action and ice–sheet glaciation may produce unimodal relationships more peaked than a first–order Fourier.

Table 4 shows first–order Fourier regressions for each of the other basic (derivative) variables as functions of aspect for a 100×100 DEM of the Torridon mountains in north–west Scotland. Stronger relationships might be found by looking at particular bands of altitude or gradient, but taking the whole area, neither convexity shows much variation with aspect. Altitude varies somewhat with aspect and gradient has a clear but still weak relationship with aspect. For both altitude and gradient the sine term is the stronger; the signs are negative and thus the maxima are south of west.

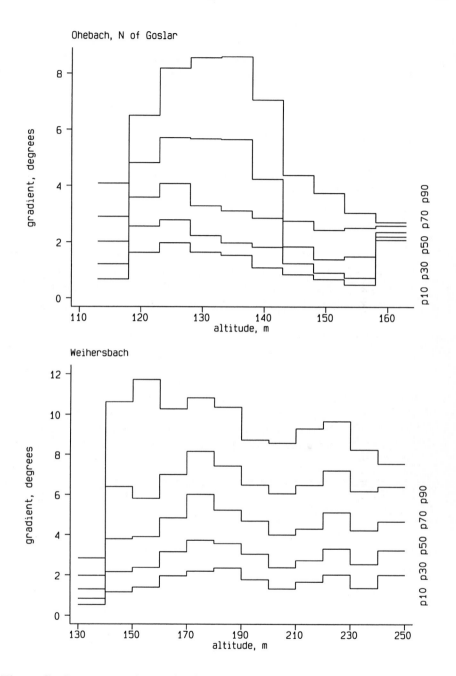

Fig. 9. Gradient versus altitude for Ohebach and Weiherbach. These plots show the common parabolic relationship, with greatest gradients at intermediate altitudes. Five percentiles of the conditional distributions are plotted for altitude classes 5 and 10 degrees wide respectively: as indicated on the right, these are the 10, 25, 75 and 90 percentiles, i.e. the median, the upper and lower quartiles, and the two outer deciles.

Table 4. Regressions on aspect, showing the effects of taking higher Fourier harmonics in predicting gradient: they are useful for Eyach but of little value for Weiherbach and Ohebach. For Eyach, focusing on altitudes below 660 m considerably increases R. s = sin(aspect), c = cos(aspect), s2 = sin(2×aspect), c2 = cos(2×aspect), etc.; R varies from 0 to 1.

	R
Torridon:	
Altitude = 449.20 − 21.79 s − 6.21 c	0.105
Gradient = 15.39 − 3.40 s − 0.77 c	0.221
Profile convexity = 0.83 + 0.347 s − 0.292 c	0.034
Plan convexity = 2.70 − 4.71 s + 0.88 c	0.053
Eyach below 660 m; 2 harmonics:	
Gradient = 15.90 − 0.82 s −1.95 c + 0.40 s2 + 1.44 c2	0.234
for 1 harmonic,	0.173;
for 3 harmonics (see Fig. 10),	0.239
over all altitudes:	
Gradient = 12.36 − 1.62 s − 0.29 c	0.146
Gradient = 12.18 −1.40 s − 0.23 c − 0.56 s2 + 0.73 c2	0.170
Wernersbach:	
Gradient = 3.64 − 0.44 s + 0.26 c	0.127
Weiherbach:	
Gradient = 5.26 − 1.42 s − 0.17 c	0.280
Gradient = 5.24 − 1.43 s − 0.16 c − 0.18 s2 + 0.18 c2 − 0.19 s3 + 0.06 c3	0.286
Sqrt (Gradient) = 2.173 − 0.268 s − 0.034 c	0.253
Ohebach:	
Gradient = 3.67 − 1.49 s − 1.04 c	0.402
Gradient = 3.57 − 1.25 s − 0.96 c − 0.29 s2 + 0.31 c2 − 0.08 s3 − 0.32 c3	0.415

For the relationship between gradient and aspect over all altitudes in the Eyach area, R = 0.17. This is increased to 0.23 for slopes below 660 m, still a very small proportion of the total variance. The first–order terms are negative, giving a south–west maximum, but the positive second–order terms give a further maximum near north (Fig. 10). The range of variation is from 13° at east to 19° at south–west: the lower part of Fig. 10 shows how the relationship changes considerably when the second harmonics are added, and slightly with the third. The first–order Fourier harmonics give R = 0.28 for Weiherbach, R = 0.13 for Wernersbach and R = 0.40 for Ohebach. Ohebach, the lowest–relief area, thus has the strongest variation of gradient with aspect. Fig. 11 shows all three harmonics for Weiherbach, giving only a slight increase in R (Table 4), but flattening the high expected gradients over a broad range of westerly aspects while leaving the minimum north of east. Wernersbach has a northwest maximum, while the other three German DEMs have southwest maxima of gradient. The first–order coefficients are stable as higher–order terms are added.

Disappointment in the weakness of relationships to aspect is based on an unrealistic expectation, for it is extremely unlikely that gradient will vary mainly with aspect given the large number of other controls such as geology and position relative to the drainage network. A relationship based on a large data set, in which predicted values of gradient vary by almost one standard deviation, is about as much as can be expected and deserves interpretation. There is a lot of scatter, but the signal is real in each of these examples. Much higher R^2 values are achieved by averaging values by aspect bands before fitting Fourier series, but we prefer to retain reference to the original data.

6   Convexity Relationships

Convexity, whether profile or plan, varies from high positive values (convexities) to high negative values (concavities). The tails of the distributions are always long (Figs. 12 and 13) and the great majority of values are near zero (straight slopes). Changes between adjacent cells may be tens of degrees in gradient and hundreds in aspect: with 10 m grid mesh, some convexities exceed ±100° /100 m in profile or ±1000° /100 m in plan. These are not necessarily data errors, although they should be checked for that possibility.

Plan convexity has no consistent relation to altitude: correlations are usually weaker than +0.2. Profile convexity is more common high in the landscape, and profile concavity is more common at low altitudes. Table 3 shows that the linear correlation between profile convexity and altitude is always positive, but usually quite weak. It rises above +0.5 only for single landforms such as cirques or drumlins, i.e. for relatively simple surfaces. For simulated fractal surfaces with fractal dimensions from 2.0 to 2.6, this correlation is between +0.07 and +0.18 (McClean 1990: 180). This rises for surfaces with higher fractal dimensions (which are unnaturally rough).

It is often assumed that profile and plan convexity vary together, with buttresses being convex in both profile and plan and hollows being concave. All

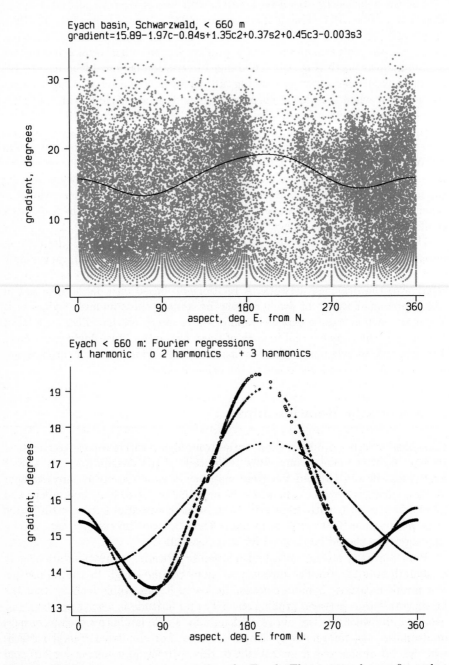

Fig. 10. Gradient versus aspect regressions for Eyach. The scatter plot confirms that festoon patterns are limited to lower gradients. The Fourier regression lines are discussed in the text.

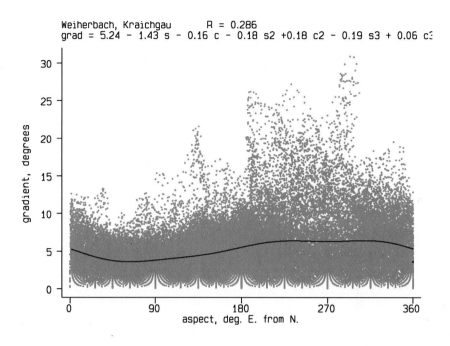

Weiherbach, Kraichgau   R = 0.286
grad = 5.24 - 1.43 s - 0.16 c - 0.18 s2 +0.18 c2 - 0.19 s3 + 0.06 c3

Fig. 11. Gradient versus aspect regressions for Weiherbach. Again, festoon patterns are seen to be limited to lower gradients.

scatter plots, however, show that this relationship is very weak, with the broad scatter and cross–like form seen in Figs. 12 and 13. The cross is slightly oblique to the axes, with more points in the ++ and the −− quadrants, so that the overall correlation is positive. Despite its low value, this correlation is remarkably consistent, between +0.142 and +0.167 here and usually between +0.10 and +0.21 for other large DEMs (Table 5). Higher correlations are provided by some individual cirques. However, the scatter plots clearly do not meet the requirements for linear correlation to be a good description of the relationship. Arctangent transformations rectify the data distributions and provide rather higher correlations; those for the Wessex area vary only between +0.20 and +0.26, and those for four DEMs in Table 5 are 45 to 80 % higher after transformation.

Finally, linear correlations between gradient and either convexity are extremely weak (below ±0.05 for the German DEMs, 0.07 for the others). Scatter plots show that there is a relationship in each case, but it is triangular, symmetrical about zero convexity (Fig. 14). Hence it is logical to disregard sign, fold over at convexity = zero, and treat convexity and concavity together. The results of these modular relationships are given in the two right–hand columns of Table 5. In all cases, the magnitude of convexity/concavity in plan declines with increasing gradient. Extreme values of plan convexity are found on floodplains and terraces: these are shown by convoluted contours and are real, not an artefact or data error. Aspect on low gradients can readily change, with a

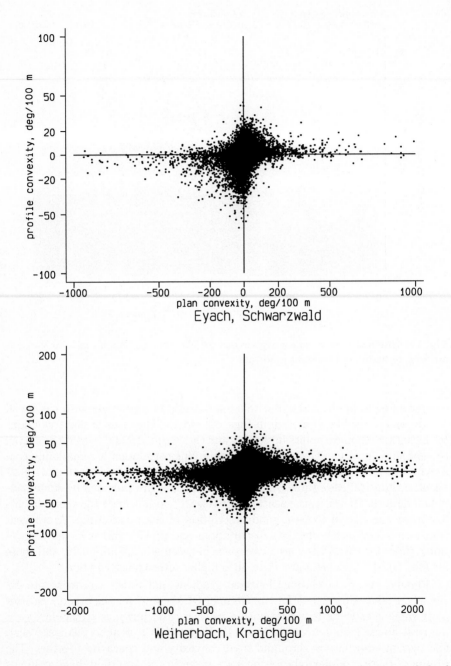

Fig. 12. Profile versus plan convexity for Eyach and Weiherbach. Scales are halved for Weiherbach. Because of strong kurtosis, especially of plan convexity, a number of points plot outside these graphs; 24 for Eyach, 135 for Weiherbach. Despite the weak correlations, the consistency of this relationship is apparent.

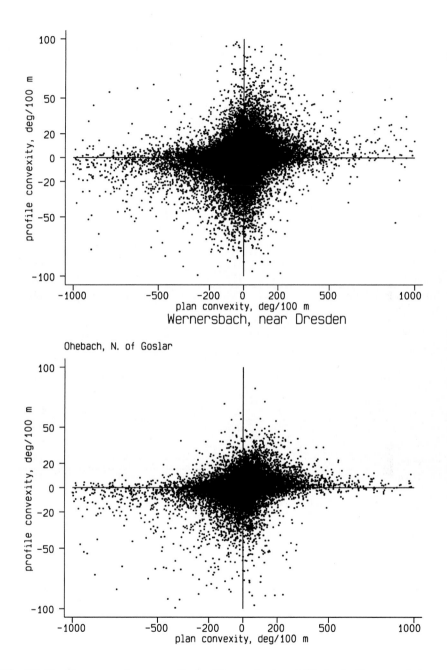

Fig. 13. Profile versus plan convexity for Wernersbach and Ohebach. Because of strong kurtosis, especially of plan convexity, a number of points plot outside these graphs; 283 for Wernersbach, and 197 for Ohebach. Despite the weak correlations, the consistency of this relationship is apparent.

Table 5. Correlations between profile and plan convexity, and gradient, for 13 European and one American DEM, and ranges of correlation for these and three further British sets of DEMs. See also Evans (1980) Table 2. Initial zeros are omitted for clarity. MOD = modulus of convexity, i.e. sign ignored; this is appropriate for triangular relationships (these correlations are weaker than .07 when signs of convexity are retained). t=arctangent transformed convexity, root–sine transformed gradient.

(a) Correlations

DEM	Grid Mesh (m)	Profile Conv. – Plan Conv.		MOD Profile Conv. – Gradient	MOD Plan Conv. – Gradient
Nupur, Iceland	100	.123		.104	−.272
Thvera, Iceland	100	.174	.308t	−.074t	−.361t
Torridon, Scotland	100	.154	.219t	.254	−.191
Rondane, Norway	100	.126		.265	−.183
Saudabotn, Norway	50	.109		−.125	−.260
Le Puy, France	50	.178		.358	−.154
Canigou, France	50	.172	.306t	−.071t	−.335t
Gardon, France	50	.169		−.477	−.347
Eyach, Germany	50	.167		.084	−.302
Uinta S., Wyoming	30	.195	.283t	.124t	−.441t
Jersey, Channel Is.	30	.098		.567	−.005
Weiherbach, Germany	12.5	.153		.277	−.259
Wernersbach, Germany	10	.142		.532	−.166
Ohebach, Germany	10	.149		.462	−.170

(b) Correlation Ranges

DEM group	Grid Mesh (m)	Profile Conv. – Plan Conv.		MOD Profile Conv. – Gradient		MOD Plan Conv. – Gradient	
		min	max	min	max	min	max
14 DEMs above	10–100	.10	.21, .31t	−.48	.57	.00	−.44
54 10x10 km; Wessex	100	.20t	.26t	.34t	.77t	−.44t	−.20t
75 Cumbrian cirques	50	.09	.47				
7 Cumbrian drumlines	50	.20	.26				

small change in the normal to the surface. All plots of plan convexity form clear triangles with concave–up sides, showing a great spread of plan convexities on low gradients.

Conversely in profile, large gradient changes do not occur in plain areas, but are feasible on steep slopes. The increase in magnitude of convexity/concavity in profile is moderately strong in Wernersbach and Ohebach and weak in the other two German DEMs. Plots are less triangular (Fig. 14, top) than those for plan convexity, and perhaps carrot–shaped. Skewness of gradient, with fewer values at high gradients, mutes the triangularity. Elsewhere, both increases and, surprisingly, decreases are found. Extreme profile convexity can combine with low gradient when the 3×3 submatrix straddles ridges or thalwegs; it would be useful to flag this situation. Relationships may be clearer after transformation of gradient and convexity.

7   Surface Properties of Areas

A limited number of studies have considered variations in these correlations, and in vector, moment or quantile statistics, over a set of DEMs. Evans (1984a,b) studied relations between 35 such statistics, and subsets of the 30, 28, 23 and 19 most useful ones, over a block of 53 contiguous 10×10 km map sheets in Wessex, a scarpland area between Bristol and Southampton. After numerous exploratory factor and principal component analyses, he concluded that at least the nine key variables listed in Table 6 needed to be retained, to avoid excessive information loss: these could be summarised in six to eight compound dimensions (factors). The nine properties listed provide distinct dimensions with some intercorrelation; they are more easily comprehended than uncorrelated (orthogonal) compound dimensions, each defined as a complex index based on the whole variable set.

For the same grid mesh and sample area size, Depraetere (1987) analysed 90 variables, including many quantiles of frequency distributions, for 72 areas providing a sample of France. His first five principal components represented 67 % of total standardised variance (Table 6, (b)); variables related to aspect were not represented in these five, but in many of the smaller components. Because as much variance as possible was allocated to the first component, this combined four of Evans' key variables. Skewness of gradient, which correlated +0.69 with skewness of altitude in the Wessex study, was separated in the French principal components. The convexity dimensions were differently expressed, but so were the variable definitions underlying them. In two studies based on smaller DEMs or profiles, Pike (1987, 1988) also demonstrated the need for numerous dimensions, including

1. steepness and altitude dispersion;
2. fine and coarse texture;
3. altitude;
4. altitude skewness;
5. curvature skewness; and
6. gradient skewness.

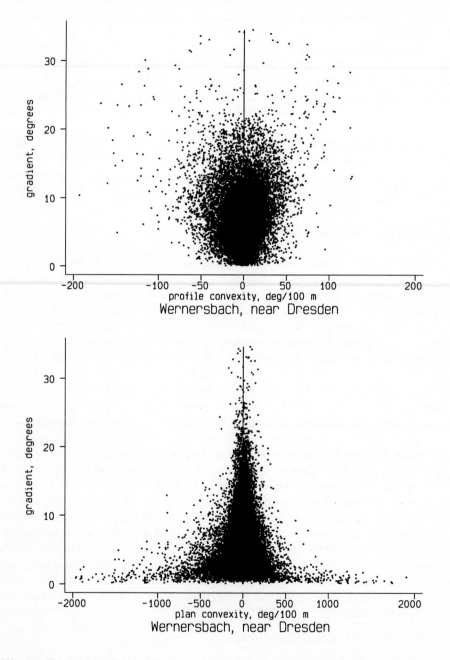

Fig. 14. Triangular relationships between convexity and gradient for Wernersbach: profile (above) and plan (below). Extreme plan convexities are confined to low gradients, on which profile convexities cluster more tightly around zero.

Table 6. Statistical dimensions of (a) the Wessex land surface, for 53 areas, each 10×10 km, after Evans 1984; and (b) the French land surface, for 72 10×10 km areas, based on Depraetere, 1987: both analysed from 50 m grids. Numbers in (b) give the order of magnitude of factors.

Property	(a) Wessex Statistical descriptor (key variable)	(b) France Dimension
Gradient	Mean gradient	1. Relief
Massiveness	Skewness of altitude	4. Skewness of altitude (& 5.)
Level	Mean altitude	* in 1.
Profile convexity	Skewness of profile convexity	2. Convexity, cols & depressions
Orientation	Weighted vector strength (modulo 180°)	–
Plan convexity	Standard deviation of plan convexity	* in 1.
Altitude–convexity	Correlation of altitude with profile convexity	3. Convexity, crests & slopes
(Profile) Variability	Standard deviation of gradient	* in 1.
Directedness	Weighted vector strength (modulo 360°)	–
		5. Skewness of gradient

Thus three scientists using different definitions of morphometric variables and working with altitude data from different areas provide a consensus that at least five independent dimensions of statistical variability are required to express the variability of real topography.

In contrast, spectral and fractal models which summarise a topography in two parameters have been very popular among engineers and physicists. Such self–affine, unifractal models are running against the grain of accumulated geomorphological knowledge. Their numerous inadequacies have been exposed by several authors, including Evans and McClean (1995); following the latter, the main divergences between such models and the real land surface are:

1. Some landforms are scale–specific
2. Landform shape varies with size
3. Variograms of altitude are curved throughout
4. Variograms of altitude differ by azimuth
5. Closed pits are less frequent than summits
6. Surface properties vary with relative height
7. Land surfaces are lineated

There are numerous other practical problems which mean that estimates of fractal dimension D are unstable and method–dependent (Klinkenberg 1994), and fail to provide useful descriptors of land surfaces. Evans and McClean concluded that unifractal or spectral models can provide only approximations which may serve as initial surfaces for simulation models, or as null hypotheses from which the deviations of real topography provide the interest.

Many geomorphologists remain convinced of scale–specificity in the landscape. This is clear in glaciated mountains (Evans and McClean 1995), in karst areas (Xiong 1992) and in aeolian dunefields (Breed and Grow 1979). In fluvially–dissected topography, drainage networks may appear self–similar across a range of scales unless environmental factors vary, but landscape dissection comes to a lower limit at the threshold between hillslope processes, with diffusive sediment transport, and channel processes. Montgomery and Dietrich (1992) showed that channeling is predictable from drainage area upslope (per unit contour width), and local gradient.

8   Conclusions

Real land surfaces show interesting deviations from random or fractal models. There are at least nine statistical dimensions of variability in regional land form: some measure vertical or horizontal scale, others involve relations between altitude, gradient and aspect which are also specific to certain regional types. Many relationships are non–linear and some are specific to an altitude or gradient band, hence it is often useful to subdivide a data set. To be applicable to the real world, geomorphological and hydrological models should be tested against a broad range of real topographies and not just on one basin or DEM.

Acknowledgments. We are grateful to all those who provided data sets, especially to Jochen Schmidt (University of Bonn) for the German data and for comments on the manuscript, and to Colin McClean (University of York) for providing some results, particularly for fractal surfaces.

References

Aandahl, A.R. (1948): The characterisation of slope positions and their influence on the total N content of a few virgin soils in Western Iowa. Soil Science Society of America, Proceedings 13: 449–454.

Barsch, D., and H. Liedtke (1980): Principles, scientific value and practical applicability of the geomorphological map of the Federal Republic of Germany at the scale of 1:25,000 (GMK 25) and 1/100,000 (GMK 100). Zeitschrift für Geomorphologie N.F. Supplement–Band 36: 296–313.

Blaszczynski, J.S. (1997): Landform characterization with GIS. Photogrammetric Engineering and Remote Sensing 63: 183–191.

Breed, C.S., and T. Grow (1979): Morphology and distribution of dunes in sand seas observed by remote sensing. In McKee, E.D. (ed.) A global study of sand seas. U.S. Geological Survey Professional Paper 1052: 253–304.

Brown, D.G., and T.J. Bara (1994): Recognition and reduction of systematic error in elevation and derivative surfaces from $7\frac{1}{2}$–Minute DEMs. Photogrammetric Engineering & Remote Sensing 60: 189–194.

Carrara, A., G. Bitelli, and R. Carla (1997): Comparison of techniques for generating DTMs from contour lines. International Journal of Geographical Information Science 11: 451–473.

Carter, J.R. (1992): The effect of data precision on the calculation of slope and aspect data using gridded DEMs. Cartographica 29: 22–34.

Chang, K., and B. Tsai (1991): The effect of DEM resolution on slope and aspect mapping. Cartography & GIS 18: 69–77.

Cox, N.J. (1997): Bin smoothing and summary on scatter plots. Stata Technical Bulletin 37: 9–12.

Cox, N.J., and A.R. Brady (1997): Spike plots for histograms, rootograms, and time–series plots. Stata Technical Bulletin 36: 8–11.

Crofts, R.S. (1981): Mapping techniques in geomorphology. In Goudie, A. et al. (eds.) Geomorphological Techniques. George Allen & Unwin, London, 66–75.

Curtis, L.F., J.C. Doornkamp, and K.J. Gregory (1965): The description of relief in field studies of soils. Journal of Soil Science 16: 16–30.

Demek, J., and C. Embleton (eds.) (1978): Guide to medium–scale geomorphological mapping. Schweizerbart'sche, Stuttgart, for International Geographical Union.

Depraetere, C. (1987): Classification automatique interrégionale à partir de MNT issus de la BDZ. Institut Géographique National, Paris, EO/DELI/SCME Rapport 4.

Evans, I.S. (1972): General geomorphometry, derivatives of altitude, and descriptive statistics. Ch.2 in Chorley, R.J. (ed.) Spatial analysis in geomorphology. Methuen, London, 17–90.

Evans, I.S. (1979): An integrated system of terrain analysis and slope mapping. Final Report on Grant DA–ERO–591–73–G0040, Department of Geography, University of Durham, England, 192 pp.

Evans, I.S. (1980): An integrated system of terrain analysis and slope mapping. Zeitschrift für Geomorphologie N.F. Supplement–Band 36: 274–295.

Evans, I.S. (1984a): Correlation structures and factor analysis in the investigation of data dimensionality: statistical properties of the Wessex land surface, England. Proceedings, International Symposium on Spatial Data Handling '84 v. 1: 98–116, Geogr. Inst., Universität Zürich–Irchel, Zürich, Switzerland.

Evans, I.S. (1984b): Properties of the Wessex land surface: an investigation of dimensionality and the choice of key variables. Durham Geomorphometry Report 8, Department of Geography, University of Durham, England, 131 pp.

Evans, I.S. (1990): General geomorphometry. Ch. 2.3 in Goudie, A. et al. (eds) Geomorphological Techniques (2nd edn), 44–56.

Evans, I.S. (1998): What do terrain statistics really mean? In Lane, S., Richards, K. and Chandler, J. (eds.) Landform monitoring, modelling and analysis. J. Wiley, Chichester, 119–138.

Evans, I.S., and C.J. McClean (1995): The land surface is not unifractal: variograms, cirque scale and allometry. Zeitschrift für Geomorphologie N.F. Suppl.–Band 101: 127–147.

Eyton, J.R. (1991): Rate–of–change maps. Cartography & GIS 18: 87–103.

Fielding, E., B. Isacks, M. Barazangi, and C.C. Duncan (1994): How flat is Tibet? Geology 22: 163–167.

Florinsky, I.V. (1996): Quantitative topographic method of fault morphology recognition. Geomorphology 16: 103–119.

Giles, P.T., and S.E. Franklin (1996): Comparison of derivative topographic surfaces of a DEM generated from stereoscopic SPOT images with field measurements. Photogrammetric Engineering & Remote Sensing 62: 1165–1171.

Guth, P.L. (1995): Slope and aspect calculations on gridded DEMs. Zeitschrift für Geomorphologie N.F. Supplement–Band 101: 31–52.

Hormann, K. (1971): Morphometrie der Erdoberfläche. Schriften des Geographischen Instituts der Universität Kiel 36: 178 pp.

Klinkenberg, B. (1994): A review of methods used to determine the fractal dimension of linear features. Mathematical Geology 26: 23–46.

Krcho, J. (1973): Morphometric analysis of relief on the basis of geometric aspects of field theory. Acta Geographica Universitatis Comenianae, Geographico–physica Nr. 1. Slovak Pedagogical Publishers, Bratislava: 7–233.

Lanyon, L.E., and G.F. Hall (1983): Land surface morphology. Soil Science 136: 291–299 & 382–386.

McClean, C.J. (1990): The scale–free and scale–bound properties of land surfaces: fractal analysis and specific geomorphometry from digital terrain models. Ph.D. thesis, University of Durham, Department of Geography: 308 pp.

Montgomery, D.R., and W.E. Dietrich (1992): Channel initiation and the problem of landscape scale. Science 255: 826–830.

Moore, I.D., R.B. Grayson, and A.R. Ladson (1991): Digital terrain modelling: a review of hydrological, geomorphological and biological applications. Hydrological Processes 5: 3–30.

Nogami, M. (1995): Geomorphometric measures for digital elevation models. Zeitschrift für Geomorphologie N.F. Supplement–Band 101: 53–67.

Pennock, D.J., B.J. Zebarth, and E. De Jong (1987): Landform classification and soil distribution in hummocky terrain, Saskatchewan, Canada. Geoderma 40: 297–315.

Petrie, G., and T.J.M. Kennie (1990): Terrain modelling in surveying and civil engineering. Whittles, Caithness.

Pike, R.J. (1987): Information content of planetary terrain: varied effectiveness of parameters for the Earth. Lunar & Planetary Science 18: 778–781.

Pike, R.J. (1988): Toward geometric signatures for geographic information systems. International Geographic Systems Symposium, Proceedings III: 15–26. NASA, Arlington.

Schmidt, J., and R. Dikau (1998): Extracting geomorphometric attributes and objects from digital elevation models – semantics, methods, future needs. In Dikau, R. and Saurer, H. (eds.) GIS in physical geography, in press.

Schmidt, J., B. Merz, and R. Dikau (1998): Morphological structure and hydrological process modelling. Zeitschrift für Geomorphologie, Supplement–Band, in press.

Shary, P.A. (1995): Land surface in gravity points classification by a complex system of curvatures. Mathematical Geology 27: 373–390.

Skidmore, A.K. (1989): A comparison of techniques for calculating gradient and aspect from a gridded DEM. International Journal of Geographical Information Systems 3: 323–334.

Speight, J.G. (1968): Parametric description of land form. In G.A. Stewart (ed.) Land evaluation. Macmillan, Melbourne, 239–250.

Summerfield, M.A. (1991): Subaerial denudation of passive margins: regional elevation versus local relief models. Earth & Planetary Science Letters 102: 406–409.

Wise, S.M. (1998): The effect of GIS interpolation errors on the use of DEMs in geomorphology. In Lane, S., Richards, K. and Chandler, J. (eds.) Landform monitoring, modelling and analysis. Chichester; J. Wiley, 139–164.

Xiong, K. (1992): Morphometry and evolution of fenglin karst in the Shuicheng area, western Guizhou, China. Zeitschrift für Geomorphologie N.F. 36: 227–248.

Young, A. (1972): Slopes. Oliver & Boyd, Edinburgh, 288pp.

Zevenbergen, L.W. and C.R. Thorne (1987): Quantitative analysis of land surface topography. Earth Surface Processes & Landforms 12: 47–56.

Zhang, W., and D.R. Montgomery (1994): DEM grid size, landscape representation, and hydrologic simulations. Water Resources Research 30: 1019–28.

Gibbs Fields with Multiple Pairwise Interactions as a Tool for Modelling Grid–Based Data

G. L. Gimel'farb[1], J. Schmidt[2], and A. Braunmandl[3]

[1] Computer Vision Unit at Tamaki, University of Auckland, New Zealand [†]
[2] Institute of Geography, University of Bonn, Germany
[3] Institute of Photogrammetry, University of Bonn, Germany

Abstract. If spatial homogeneity is restricted to only a translation in-variance then Gibbs random fields provide effective means for proba-bilistic modelling of homogeneous or piecewise–homogeneous scalar data on finite rectangular 2D grids. We discuss basic features of novel mod-els with multiple pairwise interactions between the signals in grid sites. These models show good results in simulating and segmenting piecewise–homogeneous image textures. They differ from more widely known ones, such as the autobinomial or Gauss–Markov models, in that both the in-teraction structure and strengths are learnt from a given training sample. A new learning approach, based on conditional maximum likelihood es-timates of the model parameters, provided that the training sample may rank a feasible top place in a parent population, is proposed. We applied the model to reproduce geomorphometric patterns derived from a terrain classification method. The study is aimed at showing that texture anal-ysis, based on a quantification of neighbourhood relationships, can be used to descriminate landform types. The texture segmentation shows good correlations with the terrain classification used for learning the Gibbs model parameters. The approach could be valuable in quantifying geomorphometric structures and manual terrain classification schemes.

1 Introduction

Characterization of complex topographic forms is often done through qualitative and descriptive terms, *because the nomenclature is verbal and non–unique* (Pike 1995). Fuzzy adjectives like "hilly", "rough", and "flat" are used to describe typ-ical landforms. While these qualitative taxonometric approaches offer a useful tool for geomorphic interpretation of landform and landform genesis (as dis-cussed by Dikau 1996), they suffer from the inherent subjectivity and fuzziness. The translation of lots of qualitative attributes and descriptions of landform shapes and patterns to measurable, quantitative properties offer a wide and open research field (Schmidt and Dikau 1997) and is one of the main problems in geomorphometry (Pike 1995).

[†] *on leave from* International Research and Training Center for Information Tech-nologies and Systems (National Academy of Sciences and Ministry of Education of Ukraine), Kiev, Ukraine

Recent quantitative approaches towards computational modelling of landform shapes take advantages of capabilities of GIS in geomorphometric modelling. Simple (e.g. local slope angle) and complex (e.g. upslope drainage area) geomorphometric parameters can be easily derived from digital elevation models using available GIS modules (Schmidt and Dikau 1997). Dikau (1989) uses a classification scheme to derive slope segments based on homogeneity in plan and profile curvature. Elements of a cuesta scarp can be sufficiently modeled when slope angle and upslope drainage area are introduced as additional criteria.

A multi–level landform segmentation developed by McDermid (1995) uses a multivariate statistical analysis of combined geomorphometric parameters, parameters derived by a hillslope profile analysis, and remote sensing data. The method is capable of modelling geomorphic classifications at different scales. Other methods quantify landform surfaces by using frequency statistics of geomorphometric parameters, geostatistics, network analysis, nearest neighbour analysis, spectral analysis, texture analysis, etc. (see Pike 1995).

These examples show that there are diverse approaches in geomorphometry for addressing a continuous topographic form. This implies the need for basic and systematic frameworks of geomorphometric concepts, activities and methods that can comprise, in particular, a scientific basis for geomorphometry. Pike (1995) tries to give a conceptual framework for the different approaches and activities in geomorphometry. Schmidt and Dikau (1997) present a hierarchical system for extracting geomorphometric parameters and objects from digital elevation models. However, the *landform interpretation is still an art without a formal theory* (Argialas 1995), so that there are, at least, *some basic omissions in the theoretical foundations of geomorphometry* (Schmidt and Dikau 1997). We consider three recent approaches aimed at filling this obvious gap.

One important need is to gain a more penetrating insight into understanding of the landform interpretation process. Argialas (1995) presents a prototype of a knowledge–based expert system for inferring the landforms. Such an approach aids in organizing the landform–related knowledge and can lead to more quantitative systems and models of the descriptive landform interpretation process and, moreover, to an inventory of geomorphometric forms.

Another approach is to explore the nature of the topographic form. *Reduced to its analytic essentials, topography is just geometry and topology* (Pike 1995). Several approaches towards landform classification incorporate classifications of geomorphometric features. Therefore, these methods assume that landform types can be represented by spatial homogeneity of local geomorphometric attributes. Resulting regions show a certain homogeneity in elevation, elevation range, slope angle and/or curvature. A common technique is to classify geomorphometric parameters defined at the point scale by a moving window. These methods use only simple geometric measures of the landform surface. But, the *Euclidian geometry oversimplifies so complex a surface as topography* (Pike 1995). If we look at a landscape we normally have to consider spatial patterns, formed by particular spatial relationships between the local attributes, and repetitions of these structures, e.g., valley–ridge systems. Therefore, there is a basic need for including these patterns into analysis of complex landforms. *However, topologic attributes other than those based on stream order are needed to fully describe the textures of different landscape arrangements* (Pike 1995).

The third approach is based on geomorphogenesis, that is, the origin of topographic form. If we describe the geomorphometric forms and patterns as a result of geomorphic processes acting on the landform surface, we have to consider, according to Chorley (1984), a palimpsest of the landforms resulting from different processes on different scales in time and space (Büdel 1982). Dikau (1996) discusses the history of geomorphographic landform classification and concludes that *poly–hierarchical approaches to geomorphometry which are based on nested landform hierarchies ... could deliver a conceptual framework to classify and to inventory geomorphic systems in space and time* (compare with Dikau 1989 and Dikau 1990).

Therefore, the following conclusions can be reached:

- There is a basic need for a theoretical, systematic framework for geomorphometric landform interpretation.
- Geomorphometric knowledge is mostly *descriptive and fuzzy* (Argialas 1995) and is *not systematically equated with measurable attributes* (Pike 1995). This knowledge has to be captured and, as far as possible, formulated in the explicit or quantitative terms.
- There is lack of methods and models for quantification of the geomorphometric structure and texture (Schmidt and Dikau 1997, Schmidt et al. 1997). Topology and pattern are essential elements of a complex topographic surface and, therefore, are necessary for capturing and characterizing the shape of the terrain.
- Landform surfaces have a poly–hierarchical character depending on geomorphic processes. Therefore geomorphometric analysis has to consider and to model nested landform hierarchies.

In this paper we apply a probability model which was proposed initially for analyzing image textures (Gimel'farb 1996a–1996c) to reproduce geomorphometric patterns derived from a classification method used in geomorphometry. This study is aimed at showing that texture analysis, based on a quantitative description of neighbourhood relationships, can also be used to discriminate the landform types. Because the landform segments are separated by their salient textural attributes this analysis can be directed to define landform types on the basis of training samples. By taking knowledge–based "ideal landform types" as training samples, such an approach can be useful to define and apply a geomorphometric inventory of the topographic forms.

2   Methodology

We address the probability modelling of grid–based data collected for or related to the Earth's surface provided that certain features of the surface can be expressed in terms of spatial homogeneity or piecewise homogeneity of the signals measured in the grid sites. Space and aerial images, dense digital elevation models (DEMs), or morphometric maps exemplify such data.

The homogeneity in this context means that certain signal correlations or interactions between the sites, having a specific relative arrangement, are independent, in the general case, of their absolute positions, scale, and orientation in the grid. For instance, some homogeneous landform types can be described by the numerical values of surface flatness, local elevation range, and profile type of the DEM which are computed from the elevations within a moving window (Dikau et al. 1995). In image processing such homogeneity is peculiar to the image textures. Generally, the "texture" is too fuzzy notion so that we shall restrict our consideration to a specific subset of the homogeneous image textures, called *uniform stochastic textures* by Gimel'farb (1996a), which allows for an explicit quantitative definition.

Let us assume that the homogeneity is restricted to only a translation invariance of conditional probabilities of particular signal subsets. Then *Gibbs random fields* (GRF), that is, random fields with Gibbs probability distributions (GPD), are widely used to model the homogeneous or piecewise–homogeneous data. The GRFs were first introduced in statistical physics for describing large physical systems of interacting particles (see, for instance, Isihara 1971). The models on finite grids which are most interesting for modelling images, DEMs, or other spatially distributed digital data were studied theoretically by Averintsev (1970, 1972), Dobrushin (1968, 1976), Hammersley, and Besag (1974). Hassner and Sklansky (1980), Cross and Jain (1983), Lebedev et al. (1983), Kashyap and Chellappa (1983), Derin et al. (1983) , Geman and Geman (1983) pioneered in applying the GRFs to model, enhance, and segment the noisy and textured images. Then, these models were elaborated in a wealth of researches, see, e.g., the excellent collection of papers edited by Chellappa and Jain (1993) and comprehensive surveys of Kashyap (1986), Tuceryan and Jain (1993), Li (1995), and Winkler (1995).

The well–known theorems of Averintsev (Averintsev 1972) and Hammersley and Clifford (Besag 1974) have stated that Markov random fields (MRF), under a positivity condition of a non–zero probability of each signal configuration in the grid, belong to the GRFs and that the joint and conditional GPDs are factored in such a way that the factors are specified explicitly by a geometric structure and quantitative strengths of the local signal interactions. Each factor is strictly positive and depends on the signals in a supporting subset of the grid sites. These factors are written usually in the exponential form and the exponent is called the (Gibbs) *potential*. The higher the potential value, the stronger the interaction, that is, the more probable the signal configuration that yields this value. Such a factorization ensures a mutual compatibility of the joint and conditional GPDs.

The interaction structure is given by a *neighbourhood graph*. Its edges link each pair of the interacting sites, called the neighbours. The Hammersley–Clifford theorem states that in the MRF case the factors are supported by complete subgraphs, or *cliques*, of the neighbourhood graph (Besag 1974). Some non–Markov GRFs also allow the like factorization (Gimel'farb 1996b). Here, the supports are the cliques of the subgraph representing most characteristic and stable local part of the total "grid–wide" interactions.

The GRFs are promising for modelling homogeneous and piecewise–homogeneous data on the finite grids because (*i*) the model samples, under a given GPD, are generated easily by pixel–wise stochasic relaxation techniques (Cross and Jain 1983, Geman and Geman 1983) and (*ii*) the model parameters can be learnt (estimated) from given training samples (Kashyap and Chellappa 1983, Chellappa and Jain 1993). But, the traditional Gibbs models of modern applied statistics, for instance, the auto–binomial or Gauss–Markov ones, have to a great extend pre–defined interaction structures and potentials. Thus, by and large, they may not be the best for describing the images, DEMs, or other spatial data. More novel GRFs with multiple pairwise interactions (Gimel'farb 1996a–1996c) have higher adaptability because both the interaction structure and potentials are learnt from a training sample by initial analytic and subsequent stochastic approximation of the potentials.

In this paper we discuss basic features of the GRFs with multiple pairwise interactions and study their ability in segmenting raster–based DEMs onto meaningful homogeneous textured regions. The main advantages of texture analysis by these GRFs for the landform analysis are as follows.

– The definition of homogeneity is based on neighbourhood graphs which describe the spatial relationships of geomorphometric features over the supporting grid. This could be useful, e.g., in recovering oriented valley–ridge structures.
– The method, based on learning the model parameters from the given training samples, is open to be applied for analyzing different classification schemes or definitions of landforms.
– The terrain types are aggregated by segmenting with the learnt model parameters which is independent of artefacts like sizes of the moving window or arbitrary user–defined thresholds between the classes of geomorphometric attributes (such as shown below).

We used the described texture segmentation method to analyze a comparatively simple geomorphometric classification scheme. The semi–quantitative landform classification scheme of Hammond (1964) was adapted by Dikau (1994, 1995) to be used with the DEMs within a GIS–environment. The automated scheme is based on moving–window operations and classification techniques as they are provided by most GISs. The window is moved with no overlap over the given DEM and a set of three geomorphometric attributes (measures for slope, relief and profile type) are calculated for each step (Dikau et al. 1995):

1. The area percentage of gentle slope angle, divided into four classes.
2. The local relief, divided into six classes.
3. The profile type, expressed by the area percentage of gentle slope angle located in upland or lowland, divided into four classes.

All combinations of these three attributes yield 96 possible landform sub-units for each window step, but only about half of them show a significant frequency. According to Hammond (1964), the 96 "subtypes" can be classified into 24 "types" and following 5 "main types": plains (PLA), tablelands (TAB), plains with hills and mountains (PHM), open hills and mountains (OPM), and hills and mountains (HMO). The resulting landform segmentation (see Dikau 1994 and Dikau et al. 1995) can be denotated as meso–scale landform segments. Naturally, this classification scheme strongly depends on the involved parameters (boundaries between the classes, a threshold of the gentle slope, a definition of the upland and lowland). The parameter values have to be adapted to the size of the moving window and the grid size of the given DEM. The method proposed by Hammond is a simple classification scheme taking no account of the spatial landform patterns.

Here, this method is applied to derive meso–scale landforms of two test DEMs. Then, small samples of these DEMs are used to learn the parameters of the Gibbs model of piecewise–homogeneous textures. The parameters specify a geometric structure and quantitative strengths of local pairwise interactions between the DEM data. The model with the learnt parameters is applied to obtain the landform segmentation to be compared with the initial classification.

Below, in Section 3 the mathematical grounds of the proposed approach, in particular, the homogeneous GRFs supported by the rectangular 2D grids, are considered. A uniform arc–coloured graph is defined for generalizing these supports. The learning approach (Gimel'farb 1996a) using the maximum likelihood estimates (MLE) of the Gibbs potentials is outlined. An alternative approach is proposed to reduce the number of the unknown parameters. It is based on the conditional MLE of the potentials, provided that the training sample may take on its feasible top rank in a parent population of the samples. The joint and conditional piecewise–homogeneous GRFs that possess the basic features and learning schemes akin to those of the homogeneous GRFs and allow to embed the data simulation and segmentation into the same Bayesian framework are considered. Experiments in using these tools for segmenting fragments of the DEMs onto meso–scale landform types are presented in Section 4.

3   Mathematical Basis of the Approach: Gibbs Modelling of Spatially Homogeneous and Piecewise–Homogeneous Data

3.1   Basic Notation

We use the following basic notation:

$\mathbf{x} = \{x(i) : i \in \mathbf{R};\ x(i) \in \mathbf{Q}\}$

> is a sample of scalar metric data or signals, related to a finite supporting grid $\mathbf{R}$ with the sites i;

$\mathbf{Q} = \{0, 1, ..., q_{max}\}$

is a finite set of integer signal values in the grid sites;

$$\mathbf{D} = \{-q_{max}, ..., 0, ..., q_{max}\}$$

is a finite set of pairwise signal differences;

$$\mathbf{l} = \{l(i) : i \in \mathbf{R}; \; l(i) \in \mathbf{K}\}$$

is a dense digital region map;

$$\mathbf{K} = \{0, ..., k_{max}\}$$

is a finite set of integer region labels;

$\mathbf{X}$ and $\mathbf{L}$ are parent populations of the data samples and of the region maps, respectively;

$[q_{min}(\mathbf{x}), q_{max}(\mathbf{x})]$ is the signal range for the sample $\mathbf{x}$;

$q_{min}(\mathbf{x}) = \min_{i \in \mathbf{R}}\{x(i)\}$ and $q_{max}(\mathbf{x}) = \max_{i \in \mathbf{R}}\{x(i)\}$ are the minimum and maximum signal values for the sample $\mathbf{x}$.

3.2   Non–Markov Gibbs Models of Images and DEMs

Usually, the Gibbs model of the Earth's surface data is defined on the rectangular 2D grid with the equispaced sites: $\mathbf{R} = \{i = (m, n) : m = 0, ..., M - 1; \; n = 0, ..., N - 1\}$ where (m, n) are integer 2D coordinates of the site i. Each type a of translation–invariant pairwise signal interactions in the grid is represented by a family $\mathbf{C}_a = \{(i, j) : i, j \in \mathbf{R}; \; i - j = (\mu_a, \nu_a)\}$ of the site pairs having the same relative placement. Each pair is the clique in the interaction graph for the Markov/Gibbs models (Besag 1974, Geman and Geman 1983) or in a subgraph describing most characteristic local interactions for a particular non–Markov Gibbs model (Gimel'farb 1996b). Here, $a \in \mathbf{A}$ where $\mathbf{A}$ is a set of indices of all the families. The families $\{\mathbf{C}_a : a \in \mathbf{A}\}$ show a geometric interaction structure in the rectangular grid in terms of orientations of the cliques $\varphi_a = \arctan(\mu_a/\nu_a)$ and distances $(\mu_a^2 + \nu_a^2)^{\frac{1}{2}}$ between the sites in the clique.

For the spatial data such as the images and DEMs of the Earth's surface it is natural to assume that the interaction strength is invariant to arbitrary changes of the signal ranges. Then, all the samples that differ only by the signal range are equivalent, by the Gibbs probability, to the same reference sample $\mathbf{x}^{rf}$ obtained by a signal normalization that maps their signal ranges $[q_{min}(\mathbf{x}), q_{max}(\mathbf{x})]$ onto a particular reference range, for instance, $[0, q_{max}]$.

The non–Markov Gibbs models that allow for arbitrary changes of the signal ranges are obtained by embedding the normalization $\mathbf{x} \to \mathbf{x}^{rf}$ directly into the Gibbs potentials (Gimel'farb 1996b). After this normalization, each grid site depends on all other sites in such a way that the local interactions are supplemented with a grid–wide interaction for getting the minimum $q_{min}(\mathbf{x})$ and maximum $q_{max}(\mathbf{x})$ signals. But, the latter interaction manifests itself only if the site supports the solitary minimum or maximum signal in the grid. Otherwise, only the local interactions have to be taken into account.

Generally, the non–Markov GPD with multiple pairwise interactions is as follows (for brevity, below we omit the superscript "rf" for the samples $\mathbf{x}^{\text{rf}}$):

$$\Pr(\mathbf{x}|\mathbf{V}) = Z_{\mathbf{V}}^{-1} \cdot \exp\left(\sum_{i\in\mathbf{R}} V(x(i)) + \sum_{a\in\mathbf{A}} \sum_{(i,j)\in\mathbf{C}_a} V_a(x(i), x(j))\right), \quad (1)$$

where $\mathbf{V} = \{V(q) : q \in \mathbf{Q}; \ V_a(q,q') : q,q' \in \mathbf{Q}; \ a \in \mathbf{A}\}$ denotes a centered vector of the Gibbs potential values and $Z_{\mathbf{V}}$ is a normalizing factor, or the partition function (Isihara 1971, Chellappa and Jain 1993):

$$Z_{\mathbf{V}} = \sum_{\mathbf{x}\in\mathbf{X}} \exp\left(\sum_{i\in\mathbf{R}} V(x(i)) + \sum_{a\in\mathbf{A}} \sum_{(i,j)\in\mathbf{C}_a} V_a(x(i), x(j))\right).$$

The exponent in Eq.(1) is called an (Gibbs) *energy* of the interactions.

Under an additional assumption that the second–order potentials depend only on the differences between the signals: $V_a(q,q') \equiv V_a(d = q - q')$, so that the centered potential vector is as follows: $\mathbf{V} = \{V(q) : q \in \mathbf{Q}; \ V_a(d) : d \in \mathbf{D}; \ a \in \mathbf{A}\}$, this model is simplified to the following GPD (Gimel'farb 1996b):

$$\Pr(\mathbf{x}|\mathbf{V}) = Z_{\mathbf{V}}^{-1} \cdot \exp\left(\sum_{i\in\mathbf{R}} V(x(i)) + \sum_{a\in\mathbf{A}} \sum_{(i,j)\in\mathbf{C}_a} V_a(x(i) - x(j))\right). \quad (2)$$

This GPD is represented by the equivalent exponential family distribution:

$$\Pr(\mathbf{x}|\mathbf{V}) = Z_{\mathbf{V}}^{-1} \cdot \exp(\mathbf{V} \bullet \mathbf{H}(\mathbf{x})) \equiv Z_{\mathbf{V}}^{-1} \cdot \exp(\mathbf{V} \bullet \mathbf{H}_{\text{cn}}(\mathbf{x})) \quad (3)$$

where $\mathbf{H}(\mathbf{x}) = \{H(q|\mathbf{x}) : q \in \mathbf{Q}; \ H_a(d|\mathbf{x}) : d \in \mathbf{D}; \ a \in \mathbf{A}\}$ is a vector of the signal histogram (SH) and of the signal difference histograms (SDH) with the components $H(q|\mathbf{x})$ and $H(d|\mathbf{x})$, respectively, for the reference sample $\mathbf{x}$ and $\mathbf{H}_{\text{cn}}(\mathbf{x})$ is the like vector of the centered SH and SDHs. Here and below, the subscript "cn" indicates the centering and $\bullet$ denotes the inner product. Thus, the SH and SDHs, e.g. the gray level histogram (GLH) and gray level difference histograms (GLDH) for the image, form a *sufficient statistic* for this GPD.

The *centering* of the potentials (Gimel'farb 1996a) in (1)–(3):

$$\sum_{q\in\mathbf{Q}} V(q) = 0; \quad \forall_{a\in\mathbf{A}} \sum_{q,q'\in\mathbf{Q}^2} V_a(q,q') = \sum_{d\in\mathbf{D}} V_a(d) = 0 \quad (4)$$

and the like centering of the SH and SDHs in (3) are due to the obvious relations:

$$\sum_{q\in\mathbf{Q}} H(q|\mathbf{x}) = |\mathbf{R}|; \quad \forall_{a\in\mathbf{A}} \sum_{q,q'\in\mathbf{Q}^2} H_a(q,q'|\mathbf{x}) = \sum_{d\in\mathbf{D}} H_a(d|\mathbf{x}) = |\mathbf{C}_a| \quad (5)$$

for the initial SH and SDHs. The GPDs in (1)–(3) are invariant to such a centering.

The vectors $\mathbf{V}$ and $\mathbf{H}_{cn}(\mathbf{x})$ lie in the same G-dimensional vector subspace $\mathbf{S} \subset \mathbb{R}^{G+|\mathbf{A}|+1}$ where $G = q_{max} \cdot (2|\mathbf{A}|+1)$. As shown by Barndorff-Nielsen (1978), the GPD in (3) is the regular exponential family distribution with minimal canonical parameter $\mathbf{V}$ and minimal sufficient statistic $\mathbf{H}_{cn}(\mathbf{x})$ if and only if the following conditions are both satisfied: (i) the vectors $\mathbf{V}$ are affinely independent and (ii) the vectors $\mathbf{H}_{cn}(\mathbf{x})$ are affinely independent. This holds for the models in (2) and (3) (see Appendix 1).

Thus, the GPD in (2) and (3) is strictly log–concave, that is, strongly unimodal, with respect to the potentials $\mathbf{V}$ (Barndorff–Nielsen 1978, Jacobsen 1989). As shown in Gimel'farb (1996a), this allows for learning both the interaction structure and strengths from a given training sample. We briefly outline this learning technique, as applied to the GPD in (2) and (3), below.

3.3 Learning the Model Parameters

The basic features of the non–Markov models in (1)–(3) including the parameter estimation are similar to or obtained with minor changes from the features of the Markov/Gibbs model in Gimel'farb (1996a). The learning is based on the MLE of the Gibbs potentials. It exploits the analytic first approximation of the potentials, the search for a characteristic interaction structure using the approximate potential values, and the final refinement of the potentials by a stochastic approximation.

The major distinction between the Markov and non–Markov Gibbs models is in the stochastic relaxation techniques used for generating data samples under a given GPD. Every relaxation step involves, in the Markov/Gibbs case, a summation of the potentials only over a local neighbourhood of the current site. The neighbourhood is formed by the cliques containing this site. In the non–Markov case, the local summation holds for all the sites of the reference training sample, except for the sites of the solitary maximum or minimum signal. Only in this (and rather rare) case the actual neighbourhood of the site is grid–wide so that the potentials are summed up over the total grid. Thus, the computational complexity of the relaxation is $O(|\mathbf{R}|)$ in both the cases and does not increase substantially in the non–Markov case relative to the Markov/Gibbs one.

The log–likelihood function $L(\mathbf{V}|\mathbf{x}^\circ) = \dfrac{1}{|\mathbf{R}|} \ln(\Pr(\mathbf{x}^\circ|\mathbf{V}))$ of the potential vector $\mathbf{V}$ for a training sample $\mathbf{x}^\circ$ is strictly concave and has a unique finite maximum or, in other words, the MLE of the potentials exists if and only if the following conditions hold for the marginal sample frequencies $F(q|\mathbf{x}^\circ) = H(q|\mathbf{x}^\circ)/|\mathbf{R}|$ and $F_a(d|\mathbf{x}^\circ) = H_a(d|\mathbf{x}^\circ)/|\mathbf{C}_a|$ of the signals and signal differences, respectively (see Barndorff–Nielsen 1978, Jacobsen 1989 for details):

$$\forall_{q \in \mathbf{Q}} \ 0 < F(q|\mathbf{x}^\circ) < 1; \quad \forall_{a \in \mathbf{A};\, d \in \mathbf{D}} \ 0 < F_a(d|\mathbf{x}^\circ) < 1. \tag{6}$$

The maximum of the log–likelihood function, given the centering of (4), is obtained in the point $\mathbf{V}^* \in \mathbf{S}$ where the gradient of this function is equal to zero.

The components of the gradient are as follows:

$$\frac{\partial L(\mathbf{V}|\mathbf{x}^\circ)}{\partial V(q)} = F_{cn}(q|\mathbf{x}^\circ) - \mathcal{E}\{F_{cn}(q|\mathbf{x})|\mathbf{V}\} \equiv F_{cn}(q|\mathbf{x}^\circ) - M_{cn}(q|\mathbf{V});$$

$$\frac{\partial L(\mathbf{V}|\mathbf{x}^\circ)}{\partial V_a(d)} = \rho_a \cdot (F_{cn,a}(d|\mathbf{x}^\circ) - \mathcal{E}\{F_{cn,a}(d|\mathbf{x})|\mathbf{V}\}) \tag{7}$$

$$\equiv \rho_a \cdot (F_{cn,a}(d|\mathbf{x}^\circ) - M_{cn,a}(d|\mathbf{V}))$$

so that the gradient lies in $\mathbf{S}$, too. Here, $M_{...}(\ldots|\mathbf{V}) \equiv \mathcal{E}\{\ldots|\mathbf{V}\}$ is the centered marginal probability, or the expectation of a centered marginal sample frequency, under the GPD of (3) with the potential vector $\mathbf{V}$, and the factor $\rho_a = |\mathbf{C}_a|/|\mathbf{R}|$.
The following system of equations:

$$\forall_{q\in\mathbf{Q}} \ F_{cn}(q|\mathbf{x}^\circ) = M_{cn}(q|\mathbf{V}^*); \quad \forall_{d\in\mathbf{D};\ a\in\mathbf{A}} \ F_{cn,a}(d|\mathbf{x}^\circ) = M_{cn,a}(d|\mathbf{V}^*) \tag{8}$$

holds at the unique maximum point of the likelihood function.

The samples $\mathbf{x}$, under the GPD with the known potentials $\mathbf{V}$, can be generated by pixel–wise stochastic relaxation techniques (Cross and Jain 1983, Geman and Geman 1983, Chellappa and Jain 1993). This makes possible (Younes 1988) to find the desired MLE by solving the system (8) with a stochastic approximation starting from a first approximation of the potentials.

The Analytic First Approximation of the MLE $\mathbf{V}^*$ is derived by Gimel'farb (1996a). The log–likelihood function is expanded into a truncated Taylor's series about the zero point $\mathbf{V} = \mathbf{0}$ corresponding to the independent random field (IRF). The expansion is maximized along the gradient in this point. For the GPD in (3) these estimates are as follows:

$$\forall_{q\in\mathbf{Q}} V_{[0]}(q) = \lambda_{[0]} \cdot D_{[0]}(q);$$
$$\forall_{d\in\mathbf{D};\ a\in\mathbf{A}} V_{a,[0]}(d) = \lambda_{[0]} \cdot \rho_a \cdot D_{a,[0]}(d) \tag{9}$$

where, for brevity, we denote $D_{[0]}(q) = F_{cn}(q|\mathbf{x}^\circ) - M_{cn,irf}(q)$ and $D_{a,[0]}(d) = F_{cn,a}(d|\mathbf{x}^\circ) - M_{cn,dif}(d)$. The marginal probabilities of the signals and signal differences in the IRF have the following well–known forms:

$$\forall_{q\in\mathbf{Q}} \quad M_{irf}(q) = \frac{1}{1+q_{max}}; \quad M_{cn,irf}(q) = 0;$$

$$\forall_{d\in\mathbf{D};\ a\in\mathbf{A}} M_{dif}(d) = \frac{1+q_{max}-|d|}{(1+q_{max})^2}; \quad M_{cn,dif}(d) = M_{dif} - \frac{1}{1+2q_{max}}, \tag{10}$$

and the scaling factor $\lambda_{[0]}$ is computed from the known marginals as follows:

$$\lambda_{[0]} = \frac{\sum\limits_{q\in\mathbf{Q}} D_{[0]}^2(q) + \sum\limits_{a\in\mathbf{A}} \rho_a^2 \sum\limits_{d\in\mathbf{D}} D_{a,[0]}^2(d)}{\sum\limits_{q\in\mathbf{Q}} \sigma_{irf}(q) \cdot D_{[0]}^2(q) + \sum\limits_{a\in\mathbf{A}} \rho_a^3 \sum\limits_{d\in\mathbf{D}} \sigma_{dif}(d) \cdot D_{a,[0]}^2(d)} \tag{11}$$

Here, $\sigma_{\mathrm{irf}}(q) = M_{\mathrm{irf}}(q) \cdot (1 - M_{\mathrm{irf}}(q))$ and $\sigma_{\mathrm{dif}}(d) = M_{\mathrm{dif}}(d) \cdot (1 - M_{\mathrm{dif}}(d))$.

It is worth noting that the larger the grid is, the closer the factors ρ_a are to unity, so that the estimates in (9) and (11) are almost independent of the grid size.

The GPD in (3) with $|\mathbf{A}|$ clique families involves G potential values to be estimated from $|\mathbf{R}|$ signals for a training sample. To ensure the asymptotic consistency of the MLEs, it is necessary that $G \ll |\mathbf{R}|$.

The Search for the Interaction Structure which is most characteristic for the training sample is as follows, due to simplicity of the initial potential estimates in (9). Let

$$E_{a,[0]}(\mathbf{x}^\circ) = \rho_a \sum_{d \in \mathbf{D}} \left(F_{\mathrm{cn},a}(d|\mathbf{x}^\circ) - M_{\mathrm{cn,dif}}(d) \right) \cdot F_{\mathrm{cn},a}(d|\mathbf{x}^\circ) \tag{12}$$

denote a relative Gibbs energy of the clique family a in the sample $\mathbf{x}^\circ$, given the initial potential estimates of (9), and $\mathbf{A_W}$ be the set of all the clique families covering a given large range of possible intra–clique shifts: $\mathbf{W} = \{(\mu_a, \nu_a) : |\mu_a| \leq \mu_{\max}; |\nu_a| \leq \nu_{\max}\}$.

The *interaction map* $\mathbf{E}_{[0]}(\mathbf{x}^\circ) = \{\rho_a E_{a,[0]}(\mathbf{x}^\circ) : a \in \mathbf{A_W}\}$, showing relative contributions of each clique family to the total energy, allows to recover most characteristic families comprising the desired structure. A simplest heuristic way for finding the desired characteristic families is based on a following thresholding of the interaction map:

$$\mathbf{A} = \{a : a \in \mathbf{A_W}; E_{a,[0]}(\mathbf{x}^\circ) > \theta\} \tag{13}$$

where θ denotes a given threshold. In Gimel'farb (1996a–1996c) it was chosen as a function either of the mean relative energy $\overline{E}$ and standard deviation σ_E in the interaction map: $\theta = \overline{E} + c \cdot \sigma_E$, where $c = 3 \ldots 4$, or of the maximum relative energy: $\theta = c \cdot E_{\max}$ where $c = 0.25 \ldots 0.35$. But, some further theoretical investigation is needed for optimizing such a search.

Stochastic Approximation Refinement. After finding most characteristic interaction structure, the initial estimates of (9) for the chosen clique families of (13) are refined by the stochastic approximation techniques (Younes 1988) based on a generation of a Markov chain of the model samples under a gradually changing GPD. The current potential values are updated in line with the relations between the marginal signal difference frequencies for the training sample and for the generated samples:

$$\forall_{q \in \mathbf{Q}} V_{[t]}(q) = V_{[t-1]}(q) + \lambda_{[t]} \cdot \left(F(q|\mathbf{x}^\circ) - F(q|\mathbf{x}_{[t]}) \right)$$
$$\forall_{d \in \mathbf{D}; \ a \in \mathbf{A}} V_{a,[t]}(d) = V_{a,[t-1]}(d) + \lambda_{[t]} \cdot \rho_a \cdot \left(F_a(d|\mathbf{x}^\circ) - F_a(d|\mathbf{x}_{[t]}) \right) \tag{14}$$

Here, t is the number of the approximation step ($t = 0$ for the initial estimates in (9)), $\mathbf{x}_{[t]}$ is the sample generated under the GPD $\Pr\left(\mathbf{x}|\mathbf{V}_{[t-1]}\right)$ by stochastic relaxation ($\mathbf{x}_{[0]}$ is a sample of the IRF), and the scaling factor $\lambda_{[t]}$ determines a contracted step along the current approximation of the gradient of (8).

The theoretical choice of these steps giving almost sure convergence of the updating process (14) to the desired MLE of the potentials is proposed by Younes (1988). But, such a choice affords too slow convergence to be of practical use so that a heuristic choice, introduced in Younes (1988) and slightly modified in Gimel'farb (1996a), is used in experiments presented in Gimel'farb (1996a–1996c) and in Section 4 below.

3.4   Feasible Top Rank Principle to Simplify the Learning

The foregoing learning scheme involves G unknown potential values to be computed by stochastic approximation. This number can be reduced to only $|\mathbf{A}| + 1$ unknown parameters by exploiting, instead of the unconditional MLE in (6), the conditional one provided that the training sample $\mathbf{x}$ may rank the top place attainable among all the samples of the parent population in the Gibbs energies of (12). As shown in Appendix 2, this *feasible top rank principle* produces the following conditional MLE of the potentials:

$$\forall_{q \in \mathbf{Q}} \ V^\star(q) = \lambda^\star \cdot F_{\mathrm{cn}}(q|\mathbf{x}^\circ); \quad \forall_{d \in \mathbf{D}; \ a \in \mathbf{A}} \ V_a^\star(d) = \lambda_a^\star \cdot F_{\mathrm{cn},a}(d|\mathbf{x}^\circ) \quad (15)$$

where $\lambda^\star$ are scaling factors to be computed for each clique family by maximizing the likelihood function. It is maximized in a similar way as above by approximating the factors analytically and finding the interaction structure, and then, by refining the factors for the chosen clique families with the stochastic approximation.

Generally, the estimate of (15) may differ from the true unconditional MLE of (8). But, for the GPDs in (2) some plausible considerations exist that both the estimates, at least, are fairly close if not equivalent. This conjecture needs further theoretical investigations.

The feasible top rank principle leads to the following first approximation of the factors:

$$\lambda_{[0]} = \alpha_{[0]} \cdot E_{[0]}(\mathbf{x}^\circ); \quad \forall_{a \in \mathbf{A}} \ \lambda_{a,[0]} = \alpha_{[0]} \cdot E_{a,[0]}(\mathbf{x}^\circ) \quad (16)$$

where $E_{[0]}(\mathbf{x}^\circ) = \sum_{q \in \mathbf{Q}} F_{\mathrm{cn}}^2(q|\mathbf{x}^\circ)$ is the relative first–order Gibbs energy and $E_{a,[0]}(\mathbf{x}^\circ)$ is the second–order one of (12). The factor $\alpha_{[0]}$ is computed from these energies as:

$$\alpha_{[0]} = \frac{E_{[0]}^2(\mathbf{x}^\circ) + \sum_{a \in \mathbf{A}} E_{a,[0]}^2(\mathbf{x}^\circ)}{E_{[0]}^2(\mathbf{x}^\circ) \cdot U_{[0]}(\mathbf{x}^\circ) + \sum_{a \in \mathbf{A}} E_{a,[0]}^2(\mathbf{x}^\circ) \cdot U_{a,[0]}(\mathbf{x}^\circ)}. \quad (17)$$

Here,

$$U_{[0]}(\mathbf{x}^\circ) = \sum_{q \in \mathbf{Q}} \left(F_{\mathrm{cn}}^2(q|\mathbf{x}^\circ) \cdot \sigma_{\mathrm{irf}}(q) \right)$$

$$U_{a,[0]}(\mathbf{x}^\circ) = \rho_a \sum_{d \in \mathbf{D}} \left(F_{\mathrm{cn},a}^2(d|\mathbf{x}^\circ) \cdot \sigma_{\mathrm{dif}}(d) \right) \tag{18}$$

In this case the interaction map is formed by using the weighted energies of (12): $\mathbf{E}_{[0]}(\mathbf{x}^\circ) = \left\{ \rho_a \cdot \omega_{a,[0]} \cdot E_{a,[0]}(\mathbf{x}^\circ) : a \in \mathbf{A}_\mathbf{W} \right\}$ where the weight $\omega_{a,[0]} = \sum_{d \in \mathbf{D}} F_{\mathrm{cn},a}^2(d|\mathbf{x}^\circ)$.

The stochastic approximation refinement of the factors also exploits the like energies depending on the proximity between the marginal signal difference frequencies for each clique family in the training and generated samples. At each step t of the stochastic approximation the current factors are updated as follows:

$$\lambda_{[t+1]} = \lambda_{[t]} + \alpha_{[t]} \cdot E_{[t]}(\mathbf{x}^\circ, \mathbf{x}_{[t]});$$
$$\forall_{a \in \mathbf{A}} \quad \lambda_{a,[t+1]} = \lambda_{a,[t]} + \alpha_{[t]} \cdot E_{a,[t]}(\mathbf{x}^\circ, \mathbf{x}_{[t]}) \tag{19}$$

where $\mathbf{x}_{[t]}$ is the sample generated at this step, $\alpha_{[t]}$ is the current scaling factor, and

$$E_{[t]}(\mathbf{x}^\circ, \mathbf{x}_{[t]}) = \sum_{q \in \mathbf{Q}} \left(F_{\mathrm{cn}}(q|\mathbf{x}^\circ) - F_{\mathrm{cn}}(q|\mathbf{x}_{[t]}) \right) \cdot F_{\mathrm{cn}}(q|\mathbf{x}^\circ);$$

$$E_{a,[t]}(\mathbf{x}^\circ, \mathbf{x}_{[t]}) = \rho_a \sum_{d \in \mathbf{D}} \left(F_{\mathrm{cn},a}(d|\mathbf{x}^\circ) - F_{\mathrm{cn},a}(d|\mathbf{x}_{[t]}) \right) \cdot F_{\mathrm{cn},a}(d|\mathbf{x}^\circ) \tag{20}$$

3.5 Generalized Supports

The above Gibbs models can be defined on more general finite grids. To generalize a supporting grid, let us subdivide the sites into the principal and peripheral ones depending on their features as the vertices of the interaction graph. The principal vertex (site) appears in two and only two cliques from each family, whereas the peripheral vertex belongs to a single clique or to no cliques of some types. In other words, each principal vertex joins $2 \cdot |\mathbf{A}|$ edges, two edges per clique family, but the peripheral one has less edges of some families. The complete homogeneity of the interactions holds only for the principal sites.

Let us define a *uniform edge–coloured graph* as the graph which (*i*) contains only the principal and peripheral vertices and (*ii*) possesses different colouring of its edges (in all, $|\mathbf{A}|$ colours) so that (*iii*) each principal vertex joins the same number $|\mathbf{A}|$ of the edge pairs having all the different colours and (*iv*) each peripheral vertex belongs to less number of the pairs and, possibly, to some variously coloured single edges.

These conditions can be easily checked for every particular graph. Thus, the above GRFs can be supported by any grid with arbitrary arranged sites if the involved interaction types are discriminable and form the uniform edge–coloured interaction graph.

3.6 Controllable Simulated Annealing

Usually, the stochastic approximation refinement of the potentials MLE needs a sizable number of the steps for ensuring the convergence to the desired maximum point of the likelihood function, that is, to the solution of the system (8). Thus, in practice it is sometimes difficult to implement the above learning schemes.

But, the GPDs under consideration are closely similar to the δ–function in that the images with the significantly non–zero probabilities form a very small subset concentrated around the maximum probable image(s). Therefore, the stochastic approximation processes of (14) and (19) can also be regarded as an adaptive image generating technique called in Gimel'farb (1996a, 1996c) a *controllable simulated annealing* (CSA).

The CSA generates the final images having high probabilities in relation to the GPD of (2). At each CSA step, the potentials are changing so as to approach the SH and the chosen subset of the SDHs for the training sample by the like SH and SDHs for the generated one. It differs from the usual simulated annealing (Geman and Geman 1983, Chellappa and Jain 1993) in that an explicit unimodal measure of proximity between the current generated sample and the goal training one, formed from their SHs and SDHs, is maximized.

As a result, the finally obtained samples approach more closely the training ones, as regarding their SHs and selected SDHs, than the samples that are generated by using the stochastic relaxation under the GPD with the fixed learnt potentials. In the case of image simulating, such a proximity yields, by and large, a fairly good visual similarity between the generated and goal homogeneous image textures. Thus, when using the CSA, we avoid the fairly long potential refinement stage and can simulate the model samples just after choosing the characteristic interaction structure.

3.7 Piecewise–Homogeneous Gibbs Fields

The joint GPD, describing the piecewise–homogeneous sample $\mathbf{x}$ and the corresponding map $\mathbf{l}$ of its homogeneous regions, is easily obtained from the model in (2) and (3). Each pair $(\mathbf{x}, \mathbf{l})$ is considered as a sample of the GRF such that the first– and the second–order interactions involve, respectively, a pair $(q = x(i),\ k = l(i))$ and a quadruple $(q = x(i),\ q' = x(j),\ k = l(i),\ k' = l(j))$; $(i, j) \in \mathbf{C}_a$, of the reference signals and region labels in the clique. The parent population is formed by the Cartesian product $\mathbf{X} \times \mathbf{L}$.

This model is simplified by the assumption (Gimel'farb 1996c) that the pairwise interactions depend only on the coincidence of the region labels. In other words, only two types $\alpha \in \{0, 1\}$ of the label interactions: the intra–region interaction ($\alpha = 1$, or $l(i) = l(j) = k$) and the inter–region interaction ($\alpha = 0$, or $l(i) = k \neq l(j)$) are taken into account for each region k. Then, only $(q_{max} + 1)(k_{max} + 1) + 2 \cdot (2q_{max} + 1)(k_{max} + 1)$ potential values have to be learnt per clique family.

Let $V(q, k)$ and $V_a(q, q', k, k') \equiv V_{a, \alpha = \delta(k - k')}(d = q - q', k)$ be the potentials for the pair (q, k) and the quadruple (q, q', k, k'), respectively, and $\delta(\ldots)$ denote the Kronecker function. The joint GPD is as follows:

$$\Pr(\mathbf{x}, \mathbf{l} | \mathbf{V}) = Z_{\mathbf{V}}^{-1} \cdot \exp\left(\sum_{i \in \mathbf{R}} V(x(i), l(i)) + \right.$$

$$\left. \sum_{a\in\mathbf{A}} \sum_{(i,j)\in\mathbf{C}_a} V_{a,\delta(l(i)-l(j))}(x(i)-x(j),l(i)) \right) \qquad (21)$$

and is represented by the following exponential family distribution:

$$\Pr(\mathbf{x},\mathbf{l}|\mathbf{V}) = Z_{\mathbf{V}}^{-1} \cdot \exp\left(\mathbf{V} \bullet \mathbf{H}_{\mathrm{cn}}(\mathbf{x},\mathbf{l})\right) \qquad (22)$$

where $\mathbf{V} = \{V(q,k) : (q,k) \in \mathbf{Q} \times \mathbf{K};\ V_{a,\alpha}(d,k) :\ a \in \mathbf{A};\ \alpha \in \{0,1\};\ (d,k) \in \mathbf{D} \times \mathbf{K}\}$ is the vector of the centered potentials and $\mathbf{H}_{\mathrm{cn}}(\mathbf{x},\mathbf{l}) = \{H_{\mathrm{cn}}(q,k|\mathbf{x},\mathbf{l}) :$ $(q,k) \in \mathbf{Q} \times \mathbf{K};\ H_{\mathrm{cn},a,\alpha}(d,k|\mathbf{x},\mathbf{l}): a \in \mathbf{A};\ \alpha \in \{0,1\};\ (d,k) \in \mathbf{D} \times \mathbf{K}\}$ is the vector of the centered joint histograms. Here, $H_{\mathrm{cn}}(q,k|\mathbf{x},\mathbf{l})$ are the components of the joint signal and region label histogram (S/RLH) and $H_{\mathrm{cn},a,\alpha}(d,k|\mathbf{x},\mathbf{l})$ are the components of the joint signal difference and region label coincidence histogram (SD/RLCH). These histograms are the sufficient statistic for the model. The potential centering here is quite similar to the centering of (4):

$$\sum_{k\in\mathbf{K}} \sum_{q\in\mathbf{Q}} V(q,k) = 0;\ \ \forall_{a\in\mathbf{A}}\ \sum_{k\in\mathbf{K}} \sum_{\alpha=0}^{1} \sum_{d\in\mathbf{D}} V_{a,\alpha}(d,k) = 0. \qquad (23)$$

The foregoing learning procedure holds here with obvious modifications. Notice that in this model, for simplicity, the characteristic interaction structure is assumed to be the same in all the regions. But, it is not difficult to extend the model so that each region has its own interaction structure, that is, a distinct subset $\mathbf{A}_k$ of the clique families.

The joint Gibbs model is easily reduced to conditional models by fixing either the data sample $\mathbf{x} = \mathbf{x}^\circ$ or the region map $\mathbf{l} = \mathbf{l}^\circ$. The conditional models of the data samples, given a region map $\mathbf{l}^\circ$,

$$\Pr(\mathbf{x}|\mathbf{V},\mathbf{l}^\circ) = Z_{\mathbf{V},\mathbf{l}^\circ}^{-1} \cdot \exp\left(\mathbf{V} \bullet \mathbf{H}_{\mathrm{cn}}(\mathbf{x},\mathbf{l}^\circ)\right) \qquad (24)$$

and of the region maps, given a reference data sample $\mathbf{x}^\circ$,

$$\Pr(\mathbf{l}|\mathbf{V},\mathbf{x}^\circ) = Z_{\mathbf{V},\mathbf{x}^\circ}^{-1} \cdot \exp\left(\mathbf{V} \bullet \mathbf{H}_{\mathrm{cn}}(\mathbf{x}^\circ,\mathbf{l})\right) \qquad (25)$$

differ from the joint model in (21) and (22) mainly in the parent populations ($\mathbf{X}$ and $\mathbf{L}$, respectively), partition functions, and potential centering.

In the model of (24), the homogeneous regions are fixed for all the samples $\mathbf{x} \in \mathbf{X}$. Thus, each clique family $\mathbf{C}_a$ is partitioned onto $|\mathbf{K}| = k_{\max} + 1$ fixed subfamilies $\mathbf{C}_{a,k^\circ} = \{(i,j) : (i,j) \in \mathbf{C}_a;\ l^\circ(i) = k^\circ\}$ containing each the cliques from a single region $k^\circ \in \mathbf{K}$. The potentials are centered for each subfamily individually as follows:

$$\forall_{k^\circ\in\mathbf{K};\ a\in\mathbf{A}}\ \sum_{d\in\mathbf{D}} \sum_{\alpha=0}^{1} V_{a,\alpha}(d,k^\circ) = 0. \qquad (26)$$

Generally, each region $k^\circ \in \mathbf{K}$ may have its own interaction structure $\mathbf{A}_{k^\circ}$.

Let us assume, for simplicity, that the the measurements from the different regions are mutually independent. This allows for setting to zero the potentials for all the inter–region interactions: $\forall_{k^\circ \in \mathbf{K};\, a \in \mathbf{A};\, d \in \mathbf{D}}\ V_{a,\alpha}(d, k^\circ) = 0$ and for learning the interaction structure and potentials independently in each homogeneous region of the training pair $(\mathbf{x}^\circ, \mathbf{l}^\circ)$. In this case the piecewise–homogeneous data with the known region map can be modeled by adapting the model of (2) to each type of the homogeneous data and using the learnt interaction structure(s) and Gibbs potentials in the conditional model of (24).

The model of (25) is quite symmetric to the model of (24) in that each clique family is partitioned onto $|\mathbf{D}| = 2q_{\max} + 1$ fixed subfamilies $\mathbf{C}_{a,d^\circ} = \{(i, j) : (i, j) \in \mathbf{C}_a;\ x(i) - x(j) = d^\circ\}$, $d^\circ \in \mathbf{D}$. The subfamily contains the cliques with the constant signal difference in the given data sample $\mathbf{x}^\circ$. The potential centering is as follows:

$$\forall_{d^\circ \in \mathbf{D};\, a \in \mathbf{A}}\ \sum_{k \in \mathbf{K}} \sum_{\alpha=0}^{1} V_{a,\alpha}(d^\circ, k) = 0. \tag{27}$$

In this model, generally, the characteristic interaction structure may depend on the signal differences d° so that the Gibbs energy for the pairwise interactions is computed over a union $\mathbf{A} = \bigcup_{d^\circ \in \mathbf{D}} \mathbf{A}_{d^\circ}$ of all these structures. Each clique $(i, j) \in \mathbf{A}_{d^\circ}$ is taken into account if and only if $x^\circ(i) - x^\circ(j) = d^\circ$. In this case the conditional GPD of the region maps takes the following form differing slightly from (25):

$$\Pr(\mathbf{l}|\mathbf{V}, \mathbf{x}^\circ) = Z_{\mathbf{V}, \mathbf{x}^\circ}^{-1} \cdot \exp\left(\sum_{i \in \mathbf{R}} V(x^\circ(i), l(i)) + \right. \tag{28}$$

$$\left. \sum_{d \in \mathbf{D}} \sum_{a \in \mathbf{A}_d} \sum_{(i,j) \in \mathbf{C}_a} V_{a, \delta(l(i)-l(j))}(d, l(i)) \cdot \delta\left(d - (x^\circ(i) - x^\circ(j))\right) \right)$$

The simplifying assumption that the different region labels are mutually independent allows for setting to zero the potentials for the inter–region interactions: $\forall_{d^\circ \in \mathbf{D};\, a \in \mathbf{A};\, k \in \mathbf{K}}\ V_{a,0}(d^\circ, k) = 0$ so that the potentials possess the following centering:

$$\forall_{d^\circ \in \mathbf{D};\, a \in \mathbf{A}}\ \sum_{k \in \mathbf{K}} V_{a,1}(d^\circ, k) = 0. \tag{29}$$

It is easily seen that these conditional models have the learning procedures that are quite similar to the above-mentioned one. As shown in Gimel'farb (1996c), this allows for using the same Bayesian decision framework both for simulating and segmenting piecewise–homogeneous spatial data. The foregoing CSA facilitates its practical implementation.

4   Experimental Results

Some experimental results in simulating the homogeneous or piecewise–homogeneous image textures and in segmenting piecewise–homogeneous image textures by using the proposed Gibbs models with multiple pairwise interactions are shown in Gimel'farb (1996a–1996c). Here, we restrict our consideration to the specific type of spatial data, namely, to the range images of DEMs obtained by grayscale coding of the elevations in the grid sites, or pixels, and check the abilities of the conditional model of (25) in segmenting these images into the meaningful spatially homogeneous regions. Two DEMs, one of north–central New Mexico, USA (Figure 1, compare Dikau 1994, Dikau et al. 1995) and one of Neckar catchment, southwest Germany (Figure 2), both at a spatial resolution of 200m, serve as test areas.

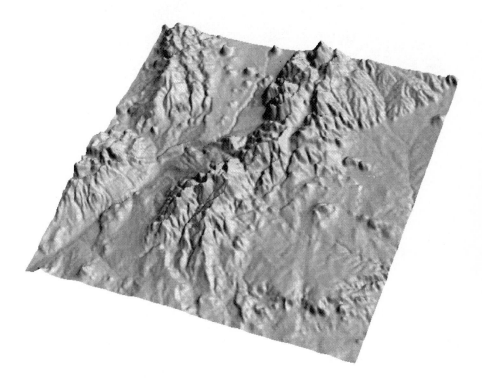

Fig. 1. Research area in New Mexico (View from SSE).

We represent these DEMs as range images (see Figures 4 and 8). For these datasets, landform segmentation maps, produced by the Hammond classification described in Section 2, are used in the experiments. The classification gives five major landform types – plains (PLA), tablelands (TAB), plains with hills and

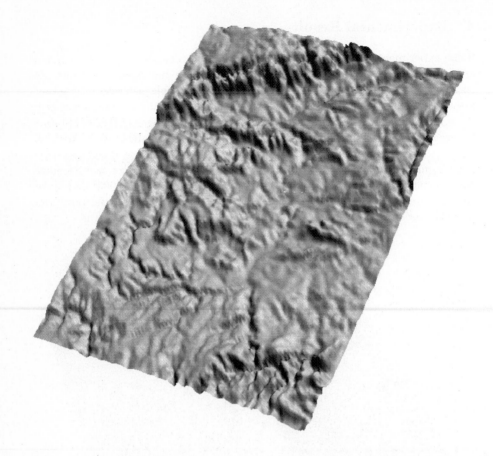

Fig. 2. Research area in the Neckar catchment (View from NNW).

mountains (PHM), open hills and mountains (OPM), and hills and mountains (HMO). Because the Gibbs model in (25) takes into account only the close– and long–range local pairwise differences between the elevations we group these five initial main types into three basic major landforms: plains and tablelands (PTL), plains with hills and mountains (PHM), open hills, hills, and mountains (OHM) (see Fig. 5 and Fig. 9). The parameters of the model were learnt from the training pairs containing small parts of the DEMs (Figures 3,a and 7,a) and their corresponding landform maps (Figures 3,b and 7,b).

Figures 3,c and 7,c show the final region map obtained by using the CSA for segmenting the training fragment of the range image with the learnt model parameters. The initial segmentation, starting from a sample of the IRF, exploits the simplified potential centering of (29). The obtained initial segmentation map, not shown here, is used as a starting point for the final segmentation based on

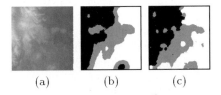

(a)   (b)   (c)

Fig. 3. Training sample pair for the DEM "New Mexico" and its segmentation map obtained with the learnt parameters of the Gibbs model. (a: the range image fragment, b: its grouped Hammond classification map and c: its segmentation by the Gibbs model).

Fig. 4. Range image of the DEM "New Mexico".

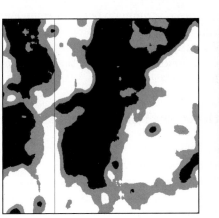

Fig. 5. Hammond classification map of the DEM "New Mexico" with white PTL, gray PHM and black OHM regions.

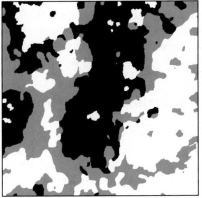

Fig. 6. Gibbs model segmentation of the DEM "New Mexico".

the potential centering of (27). The final segmentation maps give a reasonable fit to the goal training maps that validates the learnt model parameters.

The result of segmenting the DEM "New Mexico" (Figure 4) using the same learnt parameters (see Figure 3) is displayed in Figure 6. This result has to be compared with Figure 5 which shows the corresponding grouped Hammond classification map. It is evident that these maps possess rather similar macro–structures but differ in detail, especially, in the case of the intermediate PHM landform.

The equivalent results for the DEM "Neckar" are shown in Figures 7–10. It is clearly visible that they are somewhat less good than for the DEM "New Mexico". One reason might be that landform characteristics in the DEM "Neckar" are more complex on the meso–scale than the DEM "New Mexico" (see Figures 1 and 2). This results in more structured and complicated patterns of terrain segmentation.

Therefore, we attempted to use the slope data for this DEM instead of the height data. This gives a significant improvement of the obtained results, as can be seen in Figures 11–14. Here, the correspondence between the Hammond classification map and the Gibbs segmentation map is amazingly good and the differences are mainly in small details.

Because our approach of texture analysis uses only the height or slope data and no combination of geomorphometric derivates (as the approach of Hammond does), the results can be quoted as promising. Relief segmentation by using the involved Gibbs texture models can be used to discriminate between some meso–scale landforms. That means that the texture, represented here by the explicit interaction structures and strengths of multiple pairwise signal interations between the grid sites, is an essential part of geomorphometry and therefore should be useful in the terrain analysis and classification.

5 Conclusions and Implications for Further Work

The above and other similar experiments show that the homogeneous and piece-wise–homogeneous GRFs with multiple pairwise interactions provide the effective tools for analyzing the images, DEMs , and other grid–based spatial data. All these models possess a unified learning scheme involving the analytic initial and subsequent stochastic approximation of the Gibbs potentials and the search for most characteristic interaction structure and exploit a unified Bayesian processing framework based on the controllable simulated annealing.

As to the grid–based data modelling in all, the paper extends the image texture models proposed in Gimel'farb (1996a–1996c) in the following directions:

1. The models can be supported by any arbitrary grid if the involved pairwise interactions provide the uniform arc–coloured neighbourhood graph.
2. The interaction structure of the Gibbs conditional model may depend, in the general case, on the signals from a given sample to be processed.
3. The proposed principle of the feasible top rank of the training sample notably simplifies the computation of the Gibbs potential estimates.

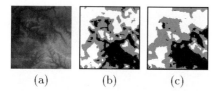

(a)   (b)   (c)

Fig. 7. Training sample pair for the DEM "Neckar" and its segmentation map obtained with the learnt parameters of the Gibbs model. (a: the range image fragment, b: its grouped Hammond classification map and c: its segmentation by the Gibbs model).

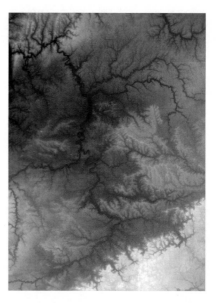

Fig. 8. Range image of the DEM "Neckar".

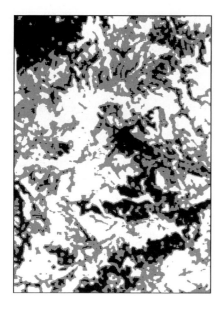

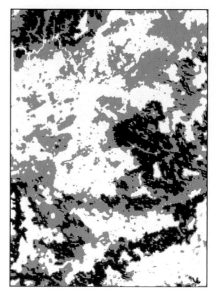

Fig. 9. Hammond classification map of the DEM "Neckar" with white PTL, gray PHM and black OHM regions.

Fig. 10. Gibbs model segmentation of the DEM "Neckar" based on the height data.

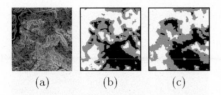

(a)   (b)   (c)

Fig. 11. Training sample pair for the DEM "Neckar" and its segmentation map obtained with the learnt parameters of the Gibbs model using the slope information. (a: the slope image fragment, b: its grouped Hammond classification map and c: its final segmentation by the Gibbs model).

Fig. 12. Map of slope angles for the DEM "Neckar".

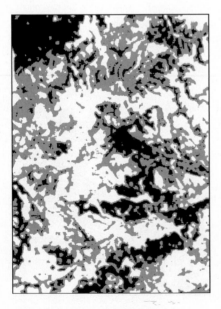

Fig. 13. Hammond classification map of the DEM "Neckar" with white PTL, gray PHM and black OHM regions.

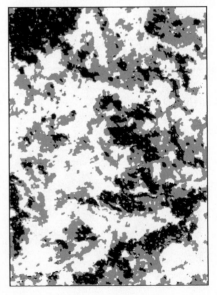

Fig. 14. Gibbs model segmentation map of the DEM "Neckar" based on the slope data.

As regards the DEMs in particular, the pairwise interactions in the elevation matrices are able for separating roughly the meso–scale landform structures that form spatially homogeneous textured regions in the range images. Our experiments show that these terrain types sometimes really form spatially homogeneous textured regions and spatial relationships between the elevations can be used as a geomorphometric feature for terrain classification. However, the efficacy of such tools for a more detailed landform discrimination remains to be investigated and some further work has to be done to develop the presented method and to make it usable for geomorphometric applications.

- Our studies show that including geomorphometric derivates other than height in the texture analysis can improve the obtained results. Therefore, it is desirable to fuse more information channels into our approach.
- Textural features (such as the pairwise interactions) and geometric features of topography should be combined to get more complete descriptions of a topographic form (Pike 1995).
- The efficiency of the Gibbs models for a more detailed landform discrimination on smaller scales (e.g., small catchments or slopes) remains to be investigated.

We expect that the method could be very helpful for investigating many approaches exploited in geomorphometry, e.g., for analyzing numerous manual and qualitative terrain classification and definition schemes. In particular, manually defined landforms could be analyzed and applied to the automatic DEM segmentation. Additionally, this method could be valuable in developing a database of geomorphometric terrain types on different scales (Argialas 1995).

References

Argialas, D.P. (1995): Towards structured–knowledge models for landform representation. Zeitschrift für Geomorphologie Suppl.–Bd. 101:85–108.

Averintsev, M.B. (1970): On one method of describing random fields with discrete argument. Problems of Information Transmission 6(2):100–108 [In Russian]

Averintsev, M.B. (1972): Description of Markov random fields using Gibbs conditional probabilities. Probability Theory and Its Applications, XVII(1):21–35. [In Russian]

Barndorff–Nielsen, O. (1978): Information and Exponential Families in Statistical Theory. Wiley.

Besag, J.E. (1974): Spatial interaction and the statistical analysis of lattice systems. J. Royal Statistical Soc. B36:192–236.

Büdel, J. (1982): Climatic Geomorphology. Princeton Univ. Press.

Chellappa, R., Jain, A. (Eds) (1993): Markov Random Fields: Theory and Application. Academic Press.

Cross, G.R. and A.K. Jain (1983): Markov random field texture models. IEEE Trans. Pattern Anal. Machine Intell. 5(1):25–39

Chorley, R.J., S.A. Schumm, and D.E. Sugden (1984): Geomorphology. Methuen.

Derin, H., H. Elliot, R. Cristi, and D. Geman (1983): Bayes smoothing algorithm for segmentation of images modelled by Markov random fields. IEEE Trans. Pattern Anal. Machine Intell. 6(6):707–720.

Dikau, R. (1989): The application of a digital relief model to landform analysis in geomorphology. In: Three–dimensional application in Geographic Information Systems (J. Raper, Ed.). Taylor&Francis, 51–77.

Dikau, R. (1990): Geomorphic landform modelling based on hierarchy theory. Proc. 4th Int. Symp. on Spatial Data Handling, July 23–27, 1990, Zürich. Vol. 1:230–239.

Dikau, R. (1994): Computergestützte Geomorphographie und ihre Anwendung in der Regionalisierung des Reliefs. Petermanns Geographische Mitteilungen 138:99–114.

Dikau, R. (1996): Geomorphologische Reliefklassifikation und –analyse. Heidelberger Geographische Arbeiten 104:15–36.

Dikau, R., E.A. Brabb, R.K. Mark, and R.J. Pike (1995): Morphometric landform analysis of New Mexico. Zeitschrift für Geomorphologie Suppl.–Bd. 101:109-126.

Dobrushin, R.L. (1968): Gibbs random fields for the lattice systems with pairwise interaction. Functional Analysis and Its Applications 2(4):31–43 [In Russian]

Dobrushin, R.L. and S.A. Pigorov (1976): Theory of random fields. Proc. 1975 IEEE–USSR Joint Workshop Information Theory, Dec. 15–19, 1975, Moscow, USSR. IEEE, 39-49.

Geman, S., and D. Geman (1983): Stochastic relaxation, Gibbs distributions, and the Bayesian restoration of images. IEEE Trans. Pattern Anal. Machine Intell. 6(6):721–741.

Gimel'farb, G.L. (1996a): Texture modeling by multiple pairwise pixel interactions. IEEE Trans. Pattern Anal. Machine Intell. 18(11):1110–1114.

Gimel'farb, G.L. (1996b): Non–Markov Gibbs texture model with multiple pairwise pixel interactions. Proc. 13th IAPR Int. Conf. Pattern Recognition, vol. II, Aug.25–29, 1996, Vienna, Austria. TUWien, 591-595

Gimel'farb, G.L. (1996c): Gibbs models for Bayesian simulation and segmentation of piecewise–uniform textures. Ibid. 760–764.

Hammond, E.H. (1964): Analysis of properties in landform geography: An application to broad–scale land form mapping. Annals of the Association of American Geographers 54:11–19.

Hassner, M. and J. Sklansky (1980): The use of Markov random fields as models of textures. Computer Graphics Image Processing 12(4):357–370.

Isihara, A. (1971): Statistical Physics. Academic Press.

Jacobsen, M. (1989): Existence and unicity of MLE in discrete exponential family distributions. Scandinav. J. Statistics 16:335–349.

Kashyap, R.L. (1986): Image models. Handbook on Pattern Recognition and Image Processing (T.Y. Young, K.–S. Fu, Eds). Academic Press, 247–279.

Kashyap, R.L. and R. Chellappa (1983): Estimation and choice of neighbours in spatial–interaction models of images. IEEE Trans. Information Theory 29(1):60–72.

Lebedev, D.S., A.A. Bezruk, and V.M. Novikov (1983): Markov Probabilistic Model of Image and Picture. Preprint: Inst. of Information Transmission Problems, Acad. Sci. USSR. VINITI. [In Russian].

Li, S.Z. (1985): Markov Random Field Modeling in Computer Vision. Springer.

McDermid, G.J. and S.E. Franklin (1995): Remote sensing and geomorphometric discrimination of slope processes. Zeitschrift für Geomorphologie Suppl.–Bd. 101:165–185.

Pike, R. (1995): Geomorphometry – progress, practice and prospect. Ibid. 221–238.

Schmidt, J. and R. Dikau (1997a): Extracting geomorphometric attributes and objects from digital elevation models – semantics, methods, future needs. GIS in Physical Geography (R. Dikau and H. Saurer, Eds), [submitted].

Schmidt, J., B. Merz, and R. Dikau (1997): Morphological structure and hydrological process modelling. Zeitschrift für Geomorphologie Suppl.–Bd., [in press].

Tuceryan, M. and A.K. Jain (1993): Texture analysis. Handbook on Pattern Recognition and Computer Vision (C.H. Chen, L.F. Pau, P.S.P. Weng, Eds). World Publishing,) 235–276.

Winkler, G. (1995): Image Analysis, Random Fields and Dynamic Monte Carlo Methods. Springer.

Younes, L. (1988): Estimation and annealing for Gibbsian fields. Annales de l'Institut Henri Poincare 24(2):269–294.

A1. Affine Independence of the Potentials and Histograms

Let $\mathbf{V} = \{V(q) : q \in \mathbf{Q}; V_a(d) : d \in \mathbf{D}; a \in \mathbf{A}\}$ be a vector of the centered potentials and $\mathbf{H}_{cn}(\mathbf{x}) = \{H_{cn}(q|\mathbf{x}) : q \in \mathbf{Q}; H_{cn,a}(d|\mathbf{x}) : d \in \mathbf{D}; a \in \mathbf{A}\}$ be a vector of the centered SH and SDHs. The affine independence of the vectors $\mathbf{V}$ is obvious because there are no restrictions on them except for the centering of (4). Thus, only the affine independence of the centered histogram vectors for the parent population $\mathbf{X}$ has to be proven. Here, we do this only for the SDHs in the histogram vectors because the proof for the SH is quite similar.

For simplicity, let us assume that the grid $\mathbf{R}$ contains, at least, q_{max} cliques of each family. To show that all the vectors $\mathbf{H}_{cn}(\mathbf{x}); \mathbf{x} \in \mathbf{X}$, are affinely independent in the subspace $\mathbf{S} \subset \mathbb{R}^{G+|\mathbf{A}|+1}$; $\dim(\mathbf{S}) = G$, of the centered vectors, let us form in this subspace $\mathbf{S}$ an orthogonal basis with G vectors. The basis vectors are stratified into $1 + |\mathbf{A}|$ groups. The first group possesses q_{max} vectors, the other ones having each $2q_{max}$ vectors. Each basis vector contains, in succession, one subvector of length $1 + q_{max}$ and $|\mathbf{A}|$ subvectors of length $1 + 2q_{max}$. All the subvectors are zero–valued except for the first subvector in the basis vectors from the first group and except for the subvector $a + 1$ in the basis vectors from the group $a + 1$ that corresponds to the second–order (pairwise) clique family a. The latter $2q_{max}$ subvectors $\{\mathbf{b}_{a,q}, \mathbf{c}_{a,q} : q = 1, \ldots, q_{max}\}$ are shown in Table 1.

Table 1. Basis subvectors.

Difference d	$-q_{max}$	$-q_{max}+1$	$-q_{max}+2$	...	-1	0	1	...	$q_{max}-2$	$q_{max}-1$	q_{max}
$\mathbf{b}_{a,1}$	-1	0	0	...	0	0	0	...	0	0	1
$\mathbf{b}_{a,2}$	0	-1	0	...	0	0	0	...	0	1	0
$\mathbf{b}_{a,3}$	0	0	-1	...	0	0	0	...	1	0	0
...	...	...	...	...	...	...	...	...	...	...	...
$\mathbf{b}_{a,q_{max}}$	0	0	0	...	-1	0	1	...	0	0	0
$\mathbf{c}_{a,1}$	-1	1	0	...	0	0	0	...	0	1	-1
$\mathbf{c}_{a,2}$	-1	-1	2	...	0	0	0	...	2	-1	-1
...	...	...	...	...	...	...	...	...	...	...	...
$\mathbf{c}_{a,q_{max}-1}$	-1	-1	-1	...	$q_{max}-1$	0	$q_{max}-1$	...	-1	-1	-1
$\mathbf{c}_{a,q_{max}}$	-1	-1	-1	...	-1	$2q_{max}$	-1	...	-1	-1	-1

The difference vectors $\mathbf{e}(\mathbf{x}, \mathbf{x}') = \mathbf{H}_{cn}(\mathbf{x}) - \mathbf{H}_{cn}(\mathbf{x}')$ for the pairs of the reference samples lie in the subspace $\mathbf{S}$, too. To prove the affine independence of the centered histogram vectors, it is sufficient to show that this holds for the difference vectors, that is, all the basis subvectors $\{\mathbf{b}_{a,q}, \mathbf{c}_{a,q} : q = 1, \ldots, q_{max}\}$ from Table 1 appear in the difference vectors for each the family a.

The samples giving the desired difference vectors are formed as follows.

(i) Let the reference sample $\mathbf{x}$ have two contiguous regions with the constant signals $q°$ and $q° + q$ in the pixels, respectively, with the exception of two pixels with the maximum (q_{max}) and minimum (0) values within the region with the signal $q°$, and let the sample $\mathbf{x}'$ have just the same regions but with the signals $q°$ and $q° - q$. The value $q°$ is chosen so that all three values $q° - q$, $q°$, and $q° + q$ are in the set $\mathbf{Q}$. Then, for $q = 1, \ldots, q_{max}$ and these sample pairs, the difference subvectors, to within a certain scaling factor, are the same as the basis subvectors $\{\mathbf{b}_{a,q} : q = 1, \ldots, q_{max}\}$.

(ii) Let the reference sample $\mathbf{x}$ contain all zero–valued signals, except for q sites with the same signal $q_{max} - q$ and one site with the signal q_{max}. The q sites are arranged in such a way that each site belongs to its "own" pair of the cliques from the family $\mathbf{C}_a$ having the signal configurations $(0, q_{max} - q)$ and $(q_{max} - q, 0)$, respectively. The sample $\mathbf{x}'$ has the same form but the signals in the above q sites possess the successive values $q_{max}, \ldots, q_{max} - q + 1$. The difference subvectors, for $q = 1, \ldots, q_{max}$, are the same in this case as the basis subvectors $\{\mathbf{c}_{a,q} : q = 1, \ldots, q_{max}\}$.

So, all the basis subvectors from Table 1 for any second–order clique family take part in the non–zero difference vectors $\mathbf{e}(\ldots)$ and hence the SDH–parts of the centered histogram vectors are affinely independent in the vector subspace $\mathbf{S}$. The independence of the SH–parts can be proven in a similar way.

A2. Feasible Top Rank Conditional MLE

For each potential vector $\mathbf{V}$, let us rank the samples $\mathbf{x} \in \mathbf{X}$ in ascending order of their total Gibbs energies of (3): $E(\mathbf{x}|\mathbf{V}) \equiv \mathbf{V} \bullet \mathbf{H}_{cn}(\mathbf{x}) = e(\mathbf{x}|\mathbf{v}) + \sum_{a \in \mathbf{A}} e_a(\mathbf{x}|\mathbf{v}_a)$

where $e(\mathbf{x}|\mathbf{v}) = \sum_{q \in \mathbf{Q}} V(q) \cdot H_{cn}(q|\mathbf{x})$ is the partial energy for the first–order clique

family (that is, for the pixel–wise cliques) and $e_a(\mathbf{x}|\mathbf{v}_a) = \sum_{d \in \mathbf{D}} V_a(d) \cdot H_{cn,a}(d|\mathbf{x})$ is the partial energy for the second–order clique family a. Here, $\mathbf{v} = \{V(q) : q \in \mathbf{Q}\}$ and $\mathbf{v}_a = \{V_a(d) : d \in \mathbf{D}\}$ denote the corresponding potential subvectors. For every clique family, the sample ranking in the partial energy is invariant to the potential (and energy) normalization that reduces the corresponding potential subvector $\mathbf{v}_{\ldots}$ to the unit subvector $\tilde{\mathbf{v}}_{\ldots} = \mathbf{v}_{\ldots}/|\mathbf{v}_{\ldots}|$.

It is easy to show that the unit subvectors $\tilde{\mathbf{v}}° = \mathbf{F}_{cn}(\mathbf{x}°)/|\mathbf{F}_{cn}(\mathbf{x}°)|$ and $\tilde{\mathbf{v}}_a° = \mathbf{F}_{cn,a}(\mathbf{x}°)/|\mathbf{F}_{cn,a}(\mathbf{x}°)|$ maximize, respectively, the normalized partial energy $e(\mathbf{x}°|\tilde{\mathbf{v}})$ and the normalized partial energy $e_a(\mathbf{x}°|\tilde{\mathbf{v}}_a)$. Here, $\mathbf{F}_{cn}(\mathbf{x}°) = \{F_{cn}(q|\mathbf{x}°) : q \in \mathbf{Q}\}$ and $\mathbf{F}_{cn,a}(\mathbf{x}°) = \{F_{cn,a}(d|\mathbf{x}°) : d \in \mathbf{D}\}$ denote, respectively, the centered vectors of the marginal signal frequencies and of the marginal signal difference frequencies for the clique family $\mathbf{C}_a$. Every arbitrary potential subvector obtained from such a unit subvector by scaling, ranks the training sample $\mathbf{x}^\circ$ in the corresponding partial energy to the same top place which may be feasible among the samples $\mathbf{x} \in \mathbf{X}$ of the parent population as compared to any other potential subvector.

The feasible top ranks for all the families lead to the following potential estimate: $\mathbf{V}^\circ(\Lambda) = \{\lambda \cdot F_{\mathrm{cn}}(q|\mathbf{x}^\circ) : q \in \mathbf{Q}; \ \lambda_a \cdot F_{\mathrm{cn},a}(d|\mathbf{x}^\circ) : d \in \mathbf{D}; \ a \in \mathbf{A}\}$. Here, $\Lambda = \{\lambda, \lambda_a : a \in \mathbf{A}; \ \lambda, \lambda_a \geq 0\}$ denotes a vector of arbitrary non–negative scaling factors.

In other words, in spite of the changes of the partial energies for the different potentials $\mathbf{V}$, the training sample may occupy the feasible top rank in the total Gibbs energy if the potentials are proportional to the centered sample marginals. The conditional MLE of the potentials $\mathbf{V}^\star = \mathbf{V}^\circ(\Lambda^\star) \equiv \{\lambda^\star \cdot F_{\mathrm{cn}}(q|\mathbf{x}^\circ) : q \in \mathbf{Q}; \ \lambda_a^\star \cdot F_{\mathrm{cn},a}(d|\mathbf{x}^\circ) : d \in \mathbf{D}; \ a \in \mathbf{A}\}$ such that $\Lambda^\star = \arg\max_{\Lambda} \Pr(\mathbf{x}^\circ|\mathbf{V}^\circ(\Lambda))$ yields the maximum probability of the training sample $\mathbf{x}^\circ$ provided that it may occupy the top rank which is feasible for this sample within the parent population ordered in the total Gibbs energies.

Adaptive Hierarchical Methods for Landscape Representation and Analysis

Th. Gerstner

Department of Applied Mathematics, University of Bonn, Germany

Abstract. Hierarchical interpolation techniques allow the efficient representation of digital terrain data. Due to the inherent adaptivity of these methods, less points are needed for the storage of smooth areas compared to non–adaptive methods. They also allow the derivation of approximate terrain models with variable level of detail. This results in a significant compression of the data as well as a highly reduced computational cost of algorithms for terrain analysis.
We will present two out of a large variety of methods, both using regular structured grids. The hierarchical triangulation method uses a successively refinable triangular grid based on right triangles. The sparse grid method, on the other hand, uses a tensor–product approach and has a better theoretical complexity. Based on these two approaches, we also construct multivariate biorthogonal wavelets using the lifting scheme. We compare all methods by means of of numerical examples.

1 Introduction

Digital elevation models (DEMs) are the starting point for many geographic applications like geographic information systems, topographic analysis, climatic, hydrological or geomorphological simulations, and landscape visualization. Most of these models are based on sampled elevation data on a rectangular grid (e. g. obtained by satellite images, radar or laser altimetry, or topographic maps) together with an interpolation method which allows the derivation of elevation values in between the grid points.

1.1 Multiresolution Models

Especially, if the grid has a small mesh size and covers a large area, the amount of data to be stored and processed can easily surmount the capabilities of today's computers. Compression algorithms try to lower or minimize storage requirements by removing redundancy in the data which is usually very prominent in DEMs. For example, in plain areas like lakes the elevation does not change, however the same elevation value is stored many times.

If the compression is lossless, the original model can be reconstructed exactly from the stored data. Otherwise, only an approximate DEM can be derived. The approximative approach usually allows much better compression rates and can be quite satisfying as long as there is some kind of control on the error caused by compression.

If the error is uniformly distributed over the whole domain, the reconstructed DEM is a representation of the original DEM at a coarser scale. In many applications such as multilevel computations it is necessary to access the data at a variety of scales. Schemes allowing several representations of the original data at various resolutions are called multiresolution models.

In this paper, we consider algorithms for the construction of such multiresolution models and for the reconstruction of approximate terrain models from these models with arbitrary accuracy. The resolution can even vary over the whole domain, allowing coarse representations in less interesting areas, finer approximations in more interesting areas and smooth transitions between the different levels of detail.

We present two methods based on adaptive hierarchical interpolation, a hierarchical triangulation method and the so–called sparse grid approach. Both methods use regular structured grids implying that the location of the grid points and their neighbour relations are fixed. Multiresolution models based on hierarchically structured grids allow fast construction and reconstruction by simple recursive algorithms. They also easily allow the parallel processing and distributed storage of the data, but their application is restricted to rectangular domains. Unstructured grid methods on the other hand are more flexible but computationally more expensive and require more memory per grid point stored.

The outline of this paper is as follows. In section 2 we will consider the hierarchical interpolation of functions in one dimension and compare the hierarchical basis with the standard nodal basis. Generalizations to two or more dimensions are presented in section 3. Here, we study the hierarchical triangulation and the sparse grid approaches in detail. We compare these methods with the help of some numerical examples. Section 4 generalizes these approaches towards biorthogonal wavelets, which can be used for scale and frequency analysis of terrains. We apply all four methods to the compression of a DEM located at the Eifel mountains. In section 5, we conclude with some remarks on further generalizations of the methods presented.

1.2 Related Work

The issue of deriving more compact representations for DEMs is a special topic of the more general problem of polygonal simplification of three–dimensional objects. It is being addressed in a variety of fields, such as computer graphics, image processing, computer aided design, computational geometry, approximation theory, and numerical mathematics.

The most popular approaches for terrain simplification are unstructured triangle meshes called TINs (triangular irregular networks), as introduced by Fowler and Little (1979). The algorithms for the construction of TINs can be roughly classified as decimation and refinement methods. In the first approach, starting from the data at full resolution, points are removed and/or relocated until the desired accuracy is reached. The latter approach starts with a simple coarse model and inserts points successively until the approximation error is less than the prescribed error threshold. For a recent survey on simplification methods see Heckbert and Garland (1997).

Multiresolution terrain models can be classified as hierarchical and pyramidal models. Hierarchical models are based on a hierarchical decomposition of the domain by recursive partitions, such as the methods we propose. Pyramidal models on the other hand are based on several approximate DEMs with different levels of detail, which are e. g. constructed by a refinement or decimation algorithm. A survey paper on multiresolution models was compiled by Puppo and Scopigno (1997).

2   Hierarchical Interpolation of Functions

The idea of hierarchical interpolation of functions is very old. In fact, it goes at least back to Archimedes who computed π approximately using a hierarchy of polygonal approximations of a circle. At the beginning of our century this approach has again been studied in approximation theory by Faber (1909). But it has not been brought to a broader attention until the eighties as an important ingredient for the efficient solution of partial differential equations (Zienkiewicz et al. 1983, Yserentant 1986).

The main idea behind hierarchical interpolation is to start with a coarse overall representation of a given function and then to improve this approximation by adding local corrections at varying scales until a certain error criterion is satisfied. We will first consider the one–dimensional case in this section, and then look at possible extensions to higher dimensions in the next section.

2.1   Problem Formulation

We assume that for a given (height) function f the input data set contains $N := 2^L + 1$ values $f(x_j), 0 \le j \le N$, on an equidistant grid $\{x_j\}$ on $\Omega = [0,1]$ with mesh width $h_L := 2^{-L}$. For simplicity of the presentation and without loss of generality we assume that $f(x_0) = f(x_N) = 0$. The interpolation problem is then to find a function $\tilde{f}$ interpolating f at the points x_j, i.e.

$$\tilde{f}(x_j) = f(x_j).$$

Usually, $\tilde{f}$ is constructed by a linear combination of (e. g. Lagrange, Hermite, spline, ...) basis functions $\{\phi_j\}$,

$$\tilde{f}(x) := \sum_{j=1}^{N-1} c_j \cdot \phi_j(x).$$

Once the basis is chosen, the coefficients c_j can be found by solving the corresponding linear system of equations

$$\mathbf{\Phi}\, \mathbf{c} = \mathbf{f},$$

where $\mathbf{\Phi} := (\phi_k(x_j))_{jk}$, $\mathbf{c} := (c_j)_j$ and $\mathbf{f} := (f(x_j))_j$.

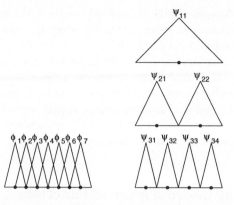

Fig. 1. Piecewise linear nodal and hierarchical basis for $L = 3$

2.2 Piecewise Linear Nodal Basis

The coefficients c_j are particularly easy to compute if a nodal (Lagrange) basis is used, i.e.

$$\phi_k(x_j) = \delta_{jk} = \begin{cases} 1 \text{ for } j = k \\ 0 \text{ for } j \neq k \end{cases}.$$

The coefficients c_j are then simply given by

$$c_j = f(x_j).$$

A simple choice for the $\{\phi_j\}$ are linear spline basis functions (Fig. 1, left) which are given by

$$\phi_j(x - x_j) := \max\left\{1 - 2^{L+1}|x - x_j|,\ 0\right\}.$$

This leads to a piecewise linear interpolant $\tilde{f}$ on the grid $\{x_j\}$.

2.3 Piecewise Linear Hierarchical Basis

The nodal basis construction does not take the smoothness of f into account. In order to accomplish this, we reorder the grid points in a certain hierarchical fashion, i.e. we assign to each grid point a level $l, 1 \leq l \leq L$. Then, we label the grid points with an index pair $(l, i), 1 \leq i \leq 2^{l-1}$, and define a bijective index transformation $\tau(l, i)$ by

$$\tau(l, i) := 2^{L-l} \cdot (2i - 1).$$

The same transformation can be applied to the basis functions, now labeled ψ_{li} (Fig.1, right). The interpolant is then constructed by

$$\tilde{f}(x) := \sum_{l=1}^{L} \sum_{i=1}^{2^{l-1}} u_{li} \cdot \psi_{li}(x) = \sum_{j=1}^{N-1} u_{\tau^{-1}(j)} \psi_{\tau^{-1}(j)}(x).$$

In order to make these coefficients efficiently computable, we use a nodal basis condition for points on the same level and we demand that the basis functions vanish on lower level points,

$$\psi_{li}(x_{kj}) = \begin{cases} \delta_{ij}, \text{ if } k = l \\ 0, \quad \text{ if } k < l. \end{cases}$$

This leads to a linear system of equations with a lower triangle matrix (if the x_{kj} are ordered with increasing k) which can be easily inverted. If we again choose a piecewise linear basis, the hierarchical linear spline basis functions $\{\psi_{li}\}$ (Fig. 1, right) are given by

$$\psi_{li}(x - x_{li}) := \max\{1 - 2^l|x - x_{li}|, \ 0\}.$$

In this case the coefficients u_{li} can even be computed in constant time and can be expressed in stencil notation by

$$u_{li} = \begin{bmatrix} -\dfrac{1}{2} & 1 & -\dfrac{1}{2} \end{bmatrix}_{x_{li}, h_l} f \ = -\frac{1}{2}f(x_{li} - h_l) + f(x_{li}) - \frac{1}{2}f(x_{li} + h_l).$$

The coefficients u_{li} can be considered as the difference between the value of f at the grid point x_{li} and the linear interpolant of the values of f from neighbouring grid points. Therefore they are also called "hierarchical surplus". Since all basis functions are piecewise linear on the grid $\{x_j\}$, this approach also leads to a piecewise linear interpolant.

2.4 Comparison of the Nodal and Hierarchical Basis

Since our interpolation problem has a unique solution, both approaches result in the same interpolant. We calculate the interpolation errors in the L_∞–norm for functions g defined by

$$\|g\|_\infty := \sup_{x \in \Omega} |g(x)|.$$

If f has the smoothness property that $f \in C^2(\Omega)$ and the L_∞–seminorm of f defined by

$$|f|_{2,\infty} := \left\| \frac{\partial^2 f}{\partial x^2} \right\|_\infty$$

is bounded, then, for the L_∞–norm of the interpolation error,

$$\|f - \tilde{f}\|_\infty = O(h_L^2),$$

holds. In this case, in contrast to the coefficients c_j, the coefficients u_{li} decrease with the level l, more exactly,

$$|u_{li}| = O(4^{-l}).$$

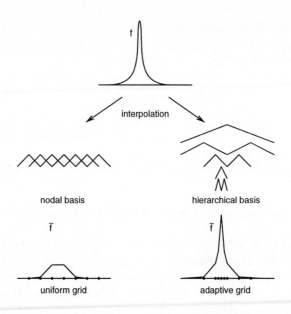

Fig. 2. Interpolation of a peak function using a nodal basis and adaptively selected hierarchical basis functions

Considering functions f which do not have the smoothness property over the whole domain, the coefficients u_{li} nevertheless get relatively small, where f is locally smooth and are relatively large, where f is not smooth. This property can be exploited by adaptive algorithms (see Fig. 2). These algorithms spend more points in areas where f is not smooth to reach a prescribed error bound with less points than non–adaptive grids do. In contrast to the nodal basis construction, the number of grid points (and therefore basis functions) can grow locally during the construction process of the interpolant.

Note that instead of the hierarchical piecewise linear basis, a hierarchical piecewise constant basis can be used, which is useful e. g. for image compression. One can also construct hierarchical polynomial bases of higher order and derive better results for the interpolation error for function classes having a higher degree of smoothness (Forsey and Bartels 1988, Bungartz 1996).

This can be used as an alternative form of adaptivity. Instead of the mesh width, the polynomial degree of the bases is considered to be variable depending on the local smoothness of f. Spatial and polynomial adaptivity can even be combined resulting in error estimates having exponential instead of polynomial decay (Babuska and Suri 1994).

2.5   Compression and Approximation

The adaptivity of the hierarchical basis can be used for data compression and approximation. For this, we first transform the whole data set into a hierarchical representation by computing the coefficients u_{li}. Then, points whose absolute

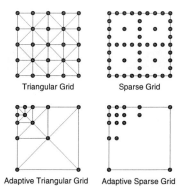

Fig. 3. Examples for regular and adaptive two–dimensional grids

coefficients $|u_{li}|$ are smaller than a prescribed cutoff value ε_{cut} are removed and only the remaining coefficients are stored. The reconstruction step takes the stored coefficients and computes the values $\tilde{f}(x_{li})$ if the corresponding coefficient u_{li} has been stored. If it has not been stored, the value $\tilde{f}(x_{li})$ is interpolated linearly from the already computed neighbouring values.

The reconstructed data is an approximation of the original data with an approximation error depending on the cutoff value ε_{cut}. The approximation error is globally not strictly bounded by ε_{cut}, because local errors can accumulate in the hierarchical representation if no smoothness is assumed for the data. In practice, however, error accumulation is quite low.

The most efficient algorithms use a bottom–up approach for data compression. They start with the points on the maximum level $l = L$ and compute the hierarchical coefficients u_{li} for each level until the lowest level $l = 1$. The reconstruction is then performed in a top–down fashion, starting with the lowest level to the maximum level. Both, compression and reconstruction can be achieved in $O(N)$ operations and can thus be applied in real time for fairly large data sets.

3   Generalizations to 2–D

There are many possibilities to generalize the one–dimensional hierarchical basis approach to two or more dimensions. For reasons of simplicity, we assume that our domain Ω is the unit square $[0, 1]^2$ with f being 0 on the boundary $\partial\Omega$. This is no restriction, because it is possible to apply the univariate method on the boundaries. The interpolant is again constructed as a linear combination of basis functions. The basis functions and interpolation stencils we will consider are suitable generalizations of their one–dimensional counterparts. We pick up two examples which both have interesting properties.

The first approach is based on a hierarchical triangulation of the domain. The corresponding grids are called triangular grids. The basis functions are pyramidal hat functions, leading to a piecewise linear interpolant. The second approach uses

Fig. 4. Pyramidal and bilinear hat function

a tensor–product subspace decomposition of the unit square which leads to so–called sparse grids. The basis functions are bilinear hat functions, leading to a piecewise bilinear interpolant (Fig. 3 and 4).

3.1   Hierarchical Triangulation

The presented hierarchical triangulation approach is commonly called "recursive bisection method" and goes back to Rivara (1984) and Mitchell (1991) who used it for the solution of elliptic partial differential equations. It is based on right triangles and is closely related to restricted quadtrees. It has been applied to terrain simplification and visualization by Gerstner (1995), Lindstrom et al. (1996), Paul and Dobler (1997) and Evans et al. (1997).

It starts with an initial triangulation of the unit square and then cuts recursively each triangle by the longest side. This approach also defines a hierarchy on the grid points being the vertices of the triangulation (Fig. 5). In this figure we have also given the grid points on the border of the domain to reveal the underlying structure of the construction despite the condition of f being zero on $\partial\Omega$. We set $\mathbf{x} = (x, y)$ and identify the grid points by their level l and a two–dimensional index $\mathbf{i}$. Furthermore, we set $\mathbf{x}_{li} =: (u, v)$. If a piecewise linear approach is taken, the basis functions are pyramidal hat functions (Fig. 4), defined for even l by

$$\psi_{li}(\mathbf{x} - \mathbf{x}_{li}) = \max\{\min\{1 - 2^{l/2}|x - u|, 1 - 2^{l/2}|y - v|\}, 0\}$$

and for odd l by a version rotated by 90 degrees,

$$\psi_{li}(\mathbf{x} - \mathbf{x}_{li}) = \max\{\min\{1 - 2^{(l-1)/2}|x - u + y - v|, 1 - 2^{(l-1)/2}|x - u - y + v|\}, 0\}.$$

The hierarchical surplus can be computed by interpolating each new grid point linearly along the side of the triangle it is placed on using the one–dimensional 3–point–stencil, eventually rotated by 90, 180, or 270 degrees (Fig. 6).

The approximation properties of this approach are well-known, e. g. from finite element analysis (Braess 1992). If $f \in C^2(\Omega)$ and all partial derivatives of f of order 2 are bounded,

$$\left\|\frac{\partial^2 f}{\partial x^i \partial y^j}\right\|_\infty < \infty, \quad \text{for } i, j \geq 0 \text{ with } i + j = 2,$$

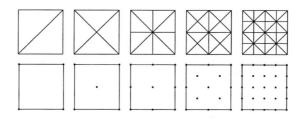

Fig. 5. Triangulation hierarchy (levels 1 to 5) and corresponding grid points

Fig. 6. Vertex dependencies in the hierarchical triangulation

then, for the approximation error

$$\|f - \tilde{f}\|_\infty = O(h_L^2)$$

holds. Here, the number of grid points is of order $O(h_L^{-2})$.

Since we have now constructed a hierarchical basis, adaptivity also is possible. But in this construction, the underlying adaptive triangulation requires a certain regularity property. It is necessary that the levels of two adjacent triangles must not differ by more than 1 (which is e.g. true in Fig. 3). If this is not the case, our representation for f using continuous basis functions is not valid and discontinuities in the approximation can arise.

3.2 Sparse Grids

The sparse grid approach goes back to Smolyak (1963), who used it for numerical integration of functions. Since then it has been applied to numerous fields such as computer graphics (Gordon 1969), interpolation (Delvos and Schempp 1989), and the solution of partial differential equations (Zenger 1991, Griebel 1991).

The main construction principle for sparse grids is a tensor product approach. Starting from our one–dimensional hierarchical basis, all possible products of basis functions are considered and combined into a two–dimensional tableau (see Fig. 7). To describe the sparse grid construction formally, we replace the single indices from the one–dimensional case by index pairs. The two–dimensional basis functions are then bilinear hat functions (Fig. 4) with varying supports in x- and y-direction, defined by

$$\psi_{\mathbf{l}\mathbf{i}}(\mathbf{x}) = \psi_{l_1 i_1}(x) \cdot \psi_{l_2 i_2}(y).$$

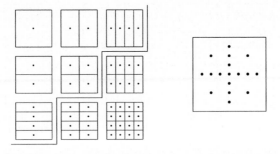

Fig. 7. Supports of the tensor–product basis functions and sparse grid

The hierarchical surplus can be computed using the analogously defined tensor product of the one–dimensional stencils,

$$u_{\mathbf{li}} = \begin{bmatrix} -\dfrac{1}{2} & 1 & -\dfrac{1}{2} \end{bmatrix}_{x_{l_1 i_1}, h_{l_1}} \otimes \begin{bmatrix} -\dfrac{1}{2} & 1 & -\dfrac{1}{2} \end{bmatrix}_{x_{l_2 i_2}, h_{l_2}} f = \begin{bmatrix} \dfrac{1}{4} & -\dfrac{1}{2} & \dfrac{1}{4} \\ -\dfrac{1}{2} & 1 & -\dfrac{1}{2} \\ \dfrac{1}{4} & -\dfrac{1}{2} & \dfrac{1}{4} \end{bmatrix}_{x_{\mathbf{li}}, h_{\mathbf{l}}} f.$$

Alternatively, the hierarchical surpluses can be computed in a two step process. In the first step, one–dimensional surpluses are computed for each row of the original data. Then, the one–dimensional stencil is applied to the just computed one–dimensional surpluses column–wise yielding the two–dimensional surpluses. Of course, the role of columns and rows can be exchanged. This procedure is computationally cheaper than using the 9–point stencil if data access is fast.

The error analysis reveals some interesting properties of this construction. Instead of the intuitive square scheme, where all basis functions with $l_1, l_2 \leq L$ are selected, a triangular scheme can be used by selecting only the basis functions above the diagonal $l_1 = l_2$ (Fig. 7). The interpolant is therefore constructed by

$$\tilde{f}(\mathbf{x}) := \sum_{l_1+l_2 \leq L+1} \sum_{i_1=1}^{2^{(l_1-1)}} \sum_{i_2=1}^{2^{(l_2-1)}} u_{\mathbf{li}} \cdot \psi_{\mathbf{li}}(\mathbf{x}).$$

The centers of the domains of the selected basis functions form a sparse grid (Fig. 7). In this case, the approximation error grows slightly compared to the full square scheme, but the number of basis functions is substantially reduced (Zenger 1991, Bungartz 1992). More exactly, if the second partial mixed derivative of f is bounded,

$$\left\| \frac{\partial^2}{\partial x^2} \frac{\partial^2}{\partial y^2} f \right\|_\infty < \infty,$$

then, for the approximation

$$\|f - \tilde{f}\|_\infty = O(h_L^2 \log h_L^{-1})$$

holds with the number of grid points only of order $O(h_L^{-1} \log h_L^{-1})$.

Since we have again formed a hierarchical basis, we can easily construct adaptive sparse grids (Fig. 3). However, unlike in the hierarchical triangulation construction, discontinuities in the interpolant can not arise.

3.3   Comparison

Both approaches can be generalized easily to arbitrary dimensions. The sparse grid approach has a better theoretical complexity, but the computation of the hierarchical surpluses is more expensive. The hierarchical triangulation method usually performs better in less smooth areas and allows the efficient graphical representation of the interpolant because of the greater locality of the basis functions.

3.4   Numerical Examples

We tested our methods on a DEM with 257×257 grid points and grid resolution 50×50 meters located in northwestern Germany at the northern border of the Eifel mountains. The maximum elevation difference is about 400 meters (Fig. 8). We computed approximative DEMs with varying accuracies between 0 and 30 meters using the adaptive hierarchical methods of the previous sections.

For the 10 and 5 meters accuracies, the sparse grid method performs slightly better than the hierarchical triangulation method. In the case of the higher accuracy DEMs, the sparse grid method requires more points than the hierarchical triangulation method (Table 1). The required grid points for the 10, 5, and 2 meter accuracies are depicted for both approaches in (Fig. 10 and 11).

In all cases, the construction of the multiresolution model and the reconstruction of the approximate DEM needed less than 0.1 seconds on a SGI O2 R10000, 150 MHz. A perspective view of a part of the area using about 15000 triangles is shown in (Fig. 9). In this picture, the error has not been uniformly distributed over the whole domain but has been weighted according to the distance to the

Table 1. Relative number of grid points used for the Eifel DEM and various accuracies with the four presented methods

	Hier. Triang.	Hier. Triang. Wav.	Sparse Grids	Sparse Grid Wav.
30 m	0.02 %	0.02 %	0.03 %	0.03 %
20 m	0.21 %	0.26 %	0.23 %	0.25 %
10 m	1.10 %	1.13 %	0.91 %	1.02 %
5 m	3.59 %	3.70 %	3.17 %	3.17 %
2 m	9.94 %	10.41 %	10.87 %	9.87 %
1 m	22.11 %	25.59 %	30.04 %	29.11 %
0 m	55.85 %	51.43 %	72.57 %	51.43 %

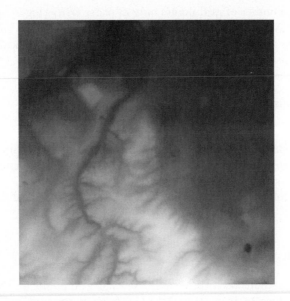

Fig. 8. Hypsoshaded DEM at the northern border of the Eifel mountains

Fig. 9. Perspective view of the DEM looking from the upper border towards the river using flat shading

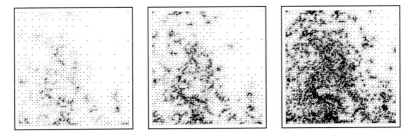

Fig. 10. Adaptive triangular grids with accuracy 10 m ($\sim$ 1.1%), 5 m ($\sim$ 3.6%) and 2 m ($\sim$ 9.9% of the original grid points) for the Eifel DEM

Fig. 11. Adaptive sparse grids with accuracy 10 m ($\sim$ 0.9%), 5 m ($\sim$ 3.2%) and 2 m ($\sim$ 10.9% of the original grid points) for the Eifel DEM

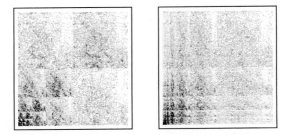

Fig. 12. Magnitude of wavelet coefficients computed by the hierarchical triangulation and the sparse grid method with lifting for the Eifel DEM

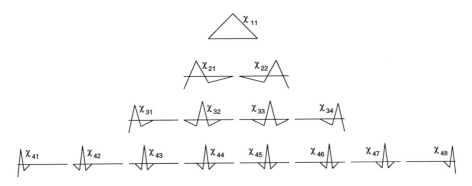

Fig. 13. Biorthogonal wavelet basis constructed by lifting for $L = 4$

viewpoint. This implies a greater level of detail near the observer and a lower level of detail further away.

In landscape visualization many approximate models with varying accuracies have to computed. In this case, it is possible to speed up computations by computing and storing additional information in each grid point and restricting the traversal of the data set to the visible parts (Paul 1997).

4 Landscape Analysis Using Wavelets

The hierarchical coefficients allow some notion about the local smoothness of a terrain. On the other hand, one is also often interested in information about local frequencies of the data. This is not possible with the help of the Fourier transformation because of its missing spatial locality. The wavelet transformation is an efficiently computable localized transformation into frequency domain (Daubechies 1992, Chui 1992). The application of wavelets and multiresolution analysis to terrain and surface simplification is described e. g. by Gross et al. (1996) and Eck et al. (1995).

With our hierarchical bases we have already taken the first step towards wavelets. Now we have to modify the construction to achieve such properties like orthogonality or vanishing moments. A univariate function g has m vanishing moments, if

$$\int g(x) \cdot x^k dx = 0$$

for $0 \leq k \leq m$. We will use a construction called "lifting" (Sweldens 1996) with piecewise linear bases, which allows us to construct biorthogonal wavelets with vanishing first moments (Cohen, et al. 1992).

The construction starts with the given input data on an equidistant grid and uses a bottom–up approach. Starting with the highest level for each level it first computes the hierarchical surplus for the grid points on the current level. But then it corrects ("lifts") the values on lower levels. This is performed in such a way that the integral of the corresponding wavelet basis functions is zero. Since the basis functions are symmetric in the interior of the domain, they will also have vanishing first moments.

We again start with the one–dimensional case. After the hierarchical ordering of the values $f_{li} := f(x_{li})$, and setting $u(x_{li}) := u_{li}$ for each level $l = L$ down to 1 the following operations are executed:

$$u_{li} := \left[-\frac{1}{2} \ \ 1 \ \ -\frac{1}{2} \right]_{x_{li}, h_l} f \qquad \text{for} \qquad 1 \leq i \leq 2^{l-1}$$

$$f_{ki} := \left[\ \ \frac{1}{4} \ \ 1 \ \ \frac{1}{4} \ \ \right]_{x_{ki}, h_l} u \qquad \text{for} \qquad 1 \leq k < l, \ 1 \leq i \leq 2^{k-1}$$

Values close to the boundary of the interval require a modification of the lifting stencil. Since one of the values u_{li} is not present for the computation of the f_{ki}, the corresponding coefficient is set to zero while the other coefficient is set to $\frac{1}{2}$. This results in the basis functions $\chi_{11} = \psi_{11}$ and for $l > 1$

$$\chi_{li}(x) := -\frac{1}{4}\psi_{\tau^{-1}(l,i)-2^{L-l}}(x) + \psi_{li}(x) - \frac{1}{4}\psi_{\tau^{-1}(l,i)+2^{L-l}}(x)$$

in the interior of the interval and with the coefficients $\frac{1}{4}$ being changed to $\frac{1}{2}$ at the boundary of the interval (Fig. 13). Note that the indices $\tau^{-1}(l,i) \pm 2^{L-l}$ refer to the neighbouring points of x_{li} on the level l.

The reconstruction is simply done top–down from level $l = 1$ to L by inverting the order of the operations and changing the sign of all fractions.

For the multivariate case, the same constructions of the previous chapter can be applied. In the sparse grid approach, the lifting stencil is also the tensor product of the one–dimensional stencil. For the hierarchical triangulation the lifting stencil has to be rotated just like the hierarchical surplus stencil, and the coefficients in the stencil are $\frac{1}{8}$ in the interior of the domain, $\frac{1}{4}$ at the edges, and $\frac{1}{2}$ at the corners. Other tensor product and triangular wavelet constructions can be found in Griebel and Oswald (1995) and Kotyczka and Oswald (1995).

In (Fig. 12) we show the wavelet coefficients for the Eifel DEM computed by the hierarchical triangulation and sparse grid methods with lifting. The wavelet coefficients are reordered hierarchically starting with the lowest level coefficients in the lower left corner to the highest level coefficients in the upper right corner of the pictures. We see that in both cases the wavelet coefficients mainly decrease with the level. High frequency peaks like the one in the lower right corner of the domain are detected more clearly by the hierarchical triangulation method. The sparse grid method, on the other hand, reveals the vertical and horizontal structure of the DEM much better.

The wavelet transformation can also be used for data compression (see section 2.5) by considering wavelet coefficients instead of hierarchical coefficients. The computing time is about twice as large as without lifting, but the compression results are just about the same (Table 1) for the DEM tested.

5   Further Extensions

There are several possible extensions to the methods presented. Instead of bi-orthogonal wavelets, one can choose semi–orthogonal or orthogonal wavelets, which have a better location in frequency domain, but are more costly to implement. There are also many other biorthogonal wavelet constructions some of which can be constructed using lifting by applying other lifting stencils, including higher order schemes which allow more vanishing moments.

In some applications it might be desirable, that certain regions of interest are approximated better than other regions, independently of the local smoothness. This can be achieved by making the cutoff value dependent on the location of the grid point considered, that is $\varepsilon_{cut} = \varepsilon_{cut}(x)$. This corresponds to a local weighting of the underlying basis functions (see also Fig. 8).

The topology of the terrain can be preserved better by using different error criteria, e.g. the difference of the normals of two adjacent triangles.

Very large data sets which do not fit into main memory can be treated by a coarse block–structuring of the data set and by loading only the currently needed blocks into memory.

So far, we have only considered the elevation data for compression. But if attributes of the terrain, like humidity or temperature are given as an attribute function over the domain Ω, they can be treated analogously. By comparing their wavelet coefficients, even correlations between attributes and the elevation data can be derived. This information can be valuable for the analysis of the data and can be used for further data compression by considering the whole data set and coding the correlations appropriately.

It is possible to construct more flexible methods by relaxing some of our initial restrictions. Without the interpolation condition $\tilde{f}(x_i) = f(x_i)$ further approximative methods can be derived by minimizing the resulting error functional. It is also possible to use a redundant basis (often called "frame" (Daubechies 1992) or "generating system" (Griebel 1994)) or even so–called "best–basis" techniques (Wickerhäuser 1996) to represent f and select basis functions from this system.

All presented methods can be generalized to arbitrary dimensions. Time–dependent data (e.g. dynamical changes of the terrain or of the humidity of the terrain) can be represented by treating time as a third dimension, if there is a finite number of time steps. For the sparse grid case, the savings compared to the full grid are then even more dramatic than in the two–dimensional case, if all mixed derivatives of order less than $2d$ are bounded. More exactly, by denoting d as the dimension, only $O(N \log N^{d-1})$ points are needed compared to $O(N^d)$ for almost the same accuracy.

References

Babuska, I., and M. Suri (1994): The p and h–p–versions of the finite element method, basic principles and properties, SIAM Rev. 36(4):578–632.

Braess, D. (1996): Finite Elemente, 2nd ed., Springer Verlag, Heidelberg.

Bungartz, H.-J. (1992): Dünne Gitter und deren Anwendung bei der adaptiven Lösung der dreidimensionalen Poisson-Gleichung, Dissertation, TU München, Institut für Informatik.

Bungartz, H.-J. (1996): Concepts for Higher Order Finite Elements on Sparse Grids, In: A.V. Ilin and L.R. Scott (eds.) Houston Journal of Mathematics: Proceedings of the 3rd Int. Conf. on Spectral and High Order Methods, Houston, 159–170.

Chui, C.K. (1992): An Introduction to Wavelets, Academic Press, Boston.

Cohen, A., I. Daubechies, and J. Feauveau (1992): Bi–orthogonal bases of compactly supported wavelets, Comm. Pure Appl. Math. 45:485–560.

Daubechies, I. (1992): Ten Lectures on Wavelets, SIAM, Philadelphia.

Delvos, F.-J., and W. Schempp (1989): Boolean methods in interpolation and approximation, Pitman Research Notes in Mathematics Series 230. Longman, Essex.

Eck, M., T. DeRose, T. Duchamp, H. Hoppe, M. Lounsbery, and W. Stuetzle (1995): Multiresolution Analysis of Arbitrary Meshes, Proc. of SIGGRAPH '95, ACM.

Evans, W., D. Kirkpatrick, and G. Townsend (1997): Right Triangular Irregular Networks, Technical Report 97–09, University of Arizona.

Faber, G. (1909): Über stetige Funktionen, Mathematische Annalen 66:81–94.

Forsey, D. and R. Bartels (1988): Hierarchical B–spline refinement, Computer Graphics, 22(4):205–212.

Fowler, R.J. and J.J. Little (1979): Automatic extraction of irregular network digital terrain models, ACM Computer Graphics 13(3):199–207.

Gerstner, T. (1995): Ein adaptives hierarchisches Verfahren zur Approximation und effizienten Visualisierung von Funktionen und seine Anwendung auf digitale 3–D Höhenmodelle, Master's thesis, TU München, Institut für Informatik.

Gordon, W.J. (1969): Distributive lattices and the approximation of multivariate functions, in Approximation with Special Emphasis on Spline Functions, I.J. Schoenberg (ed.), 223–277, Academic Press, New York.

Griebel, M. (1991): A parallelizable and vectorizable multi–level algorithm on sparse grids, In Hackbusch, W. (ed.): Parallel Algorithms for Partial Differential Equations, Notes on Numerical Fluid Mechanics 31, Vieweg, Braunschweig.

Griebel, M. (1994): Multilevel methods considered as iterative methods on semidefinite systems, SIAM Int. J. Sci. Stat. Comput. 15(3):547–565.

Griebel, M. and P. Oswald (1995): Tensor–product–type subspace splittings and multilevel iterative methods for anisotropic problems, Adv. Comput. Math. 4:171–206.

Gross, M.H., O.G. Staadt, and R. Gatti (1996): Efficient Triangular Surface Approximations using Wavelets and Quadtree Data Structures, IEEE Trans. on Visualization and Computer Graphics 2(2).

Heckbert, P.S., and M. Garland (1997): Survey of Surface Approximation Algorithms, Carnegie Mellon University Technical Report CMU–CS–97–, to appear.

Kotyczka, U., and P. Oswald (1995): Piecewise linear prewavelets of small support, in Approximation Theory VIII, Ch.K.Chui, L.L.Schumaker (eds.), vol. 2, 235–242, World Scientific Publishers.

Lindstrom, P., D. Koller, W. Ribarsky, L.F. Hodges, N. Faust, and G. Turner (1996): Real–Time, Continuous Level of Detail Rendering of Height Fields, Comp. Graph.

Mayer, S. (1992): Minimierung und Visualisierung ortsmodifizierter dünner Gitter, Master's thesis, TU München, Institut für Informatik.

Mitchell, W.F. (1991): Adaptive refinement for arbitrary finite element spaces with hierarchical bases, J. Comp. Appl. Math. 36:65–78.

Paul, A., and K. Dobler (1997): Adaptive Realtime Terrain Triangulation, Proceedings of the WSCG '97, University of West Bohemia, Plzen, Czech Republic.

Puppo, E. (1996): Variable resolution triangulations, Technical Report, 12–96, Istituto per la Matematica Applicata, C.N.R., Genova.

Puppo, E., and R. Scopigno (1997): Simplification, LOD and Multiresolution Principles and Applications. Eurographics '97 Tutorial Notes, Eurographics Association, Aire–la–Ville (CH).

Rivara, M.C. (1984): Algorithms for Refining Triangular Grids Suitable for Adaptive and Multigrid Techniques, International Journal for Numerical Methods in Engineering 20:745–756.

Smolyak, S.A. (1963): Quadrature and interpolation formulas for tensor products of certain classes of functions, Dokl. Akad. Nauk SSSR 4:240–243.

Sweldens, W. (1996): The lifting scheme: A custom–design construction of biorthogonal wavelets, Appl. Comput. Harmon. Anal. 3:186–200.

Wickerhäuser, M.V. (1996): Adaptive Wavelet–Analysis – Theorie und Software, Vieweg, Braunschweig.

Yserentant, H. (1986): On the multi–level splitting of finite element spaces, Numerische Mathematik 49:379–412.

Zenger, C. (1991): Sparse grids, In Hackbusch, W. (ed.): Parallel Algorithms for Partial Differential Equations, Notes on Num. Fluid Mech. 31, Vieweg, Braunschweig.

Zienkiewicz, O.C., J.P. Gago, and D.W. Kelly (1983): The Hierarchic Concept in Finite Element Analysis Comput. & Structures 16:53–65.

Part II

Short Term Modelling

Numerical Simulation of Surface Runoff and Infiltration of Water

G. Paul, St. Hergarten and H. J. Neugebauer

Geodynamics – Physics of the Lithosphere, University of Bonn, Germany

Abstract. The surface runoff plays a dominant part in soil erosion during excessive rainfalls. Based on the Manning formula, we present an approach that is applicable to small scales of some centimeters. Thus, structures like rills and hollows can be resolved. The infiltration is taken into account using the Richards equation. The resulting coupled differential equation system is solved numerically.

Introduction

In the field of process modelling there are several physically based formulations for describing soil erosion processes, e.g. CREAMS (Knisel 1980), WEPP (Lane and Nearing 1989), KINEROS (Woolhiser et al. 1990), EROSION2D/3D (Schmidt 1991, von Werner 1996), OPUS (Smith 1992), EUROSEM (Morgan 1989), and LISEM (de Roo et al. 1994). These models describe the surface runoff in a large scale and are able to give some good predictions in this scale. However there are also some small scaled structures like rills or hollows in the range of some centimeters. Such structures can affect the surface runoff dominantly. This is clearly recognizable in the model of Zhang and Cundy (1989) that works on a small scale. In order to investigate this affect we are dealing with a small scaled modelling of surface runoff on an arbitrary relief, too.

In general, the draining of the water is described by the Navier–Stokes equation. The solution of this equation causes a lot of numerical effort. Anyway we consider a moving boundary problem due to the fact that the surface of the water is moving. Thereby the solution of the equation is made more difficult additionally. Thus, in hydraulics the shallow water equation, which follows from a reduction of the Navier–Stokes equation, is often used. But the common shallow water formulation is not suitable for the solution of this problem, too, since the whole draining water sheet (and not only the soil boundary layer) is affected by turbulence. Therefore we consider the Manning formula (Giles 1962) that comes from the channel building and is used in many other models, too. We extend this formula for the surface runoff on an arbitrary relief.

To take the infiltration into account we consult the Richards equation (Richards 1931), which describes the water flow in a saturated or unsaturated soil.

The Manning Formula

Originally, the Manning formula (Giles 1962) was found empirically. It describes the mean velocity v of the water flow in an arbitrary channel:

$$v = \frac{1}{n} r^{\frac{2}{3}} \Delta^{\frac{1}{2}} \tag{1}$$

with:

$$\Delta = \text{slope}$$
$$r = \text{hydraulic radius} = \frac{\text{cross section}}{\text{wetted perimeter}}$$
$$n = \text{coefficient of roughness}$$

Application to Surface Runoff

Experimental results of Emmett (1970) and Pearce (1976) have shown that turbulent sheet flow on a surface can be interpreted as a flow within a wide and shallow channel. The hydraulic radius r converges to the average thickness of the water layer δ, so that Manning's equation turns into:

$$v = \frac{1}{n} \delta^{\frac{2}{3}} \Delta^{\frac{1}{2}}. \tag{2}$$

Field measurements (Weltz et al. 1992, Abrahams et al. 1994) indicated that n may vary between $0.01 \, \mathrm{m}^{-\frac{1}{3}}\mathrm{s}$ for bare sandy soils and about $1 \, \mathrm{m}^{-\frac{1}{3}}\mathrm{s}$ for shrublands or grasslands.

Extension to Small Scales and Two Dimensions

In most soil erosion models the Manning formula (2) is applied to one–dimensional slope profiles. Although this approach may be feasible for practical applications, some problems cannot be denied:

- No slope is completely homogeneous perpendicular to its gradient. The way surface runoff takes, und thus the pattern formed by erosion, is very sensitive to small inhomogeneities. The problem of averaging over these inhomogeneities has not been solved satisfactorily yet.
- Field measurements (Govers 1992; Abrahams et al. 1996) as well as geometrical investigations (Hergarten and Neugebauer 1997) have shown that the sheet flow expression (2) is not valid if there are rills. The δ–dependency may deviate from a power law significantly, and even if a power law is assumed to be valid over some range, the exponent decreases clearly. The transition between sheet flow and rill flow, occuring during strong erosion events, is not clear at all.

One way out of these problems consists of a small scale formulation in two dimensions, where at least large rills are resolved. Thus, the characteristic length scale of our approach is some centimeters. Another advantage of this approach is that the surface shape is not restricted to slopes. Naturally, this required resolution actually limits the model area to some tens of square meters.

First, we define an average flow depth $\delta(x, t)$. The two dimensional coordinate x concerns the scale resolved by the model, the averaging shall cover the smaller scales, i. e., the soil roughness.

The Manning formula (2) only concerns the absolute value of the velocity, not its direction. The most simple assumption concerning the direction is a flow downslope at each point of the surface. As it can easily be seen, this assumption causes some problems, e. g. within rills. Due to the surface gradient at the side of the rill, there will be an unrealistic permanent flow towards the middle of the rill. Even a more striking problem occurs if there is a hollow that is filled with water. Due to the surface gradient, the water flows permanently into the hollow, even if the water level within the hollow exceeds the water level outside.

These problems can be solved by taking a look at the meaning of the slope Δ in the original Manning formula (1). Here, Δ is not the slope of the channel bed, but the slope of the water level, called the energy line. Thus, we use the slope of the water level, defined by $H(x, t) + \delta(x, t)$ to determine absolute value and direction of the velocity:

$$v = -\frac{1}{n} \delta^{\frac{2}{3}} |\nabla(H + \delta)|^{\frac{1}{2}} \frac{\nabla(H + \delta)}{|\nabla(H + \delta)|} \quad . \tag{3}$$

For the sake of clarity, we have omitted the arguments x and t. Equation (3) corresponds to the two-dimensional extension of the diffusive wave approximation (Govindaraju et al. 1988, Morris and Woolhiser 1980).

The Rill as an Analytical Test

The original Manning formula was evolved for channels, and thus can be applied to turbulent flow within a rill. On the other hand, if we treat a rill using our approach, we should obtain a comparable average velocity.

In order to test this, we consider a v–shaped rill running downslope (Fig. 1). As it can be calculated analytically, our formula (2) yields an average velocity of

$$v_a = \frac{3}{4} \frac{1}{n} h^{\frac{2}{3}} \Delta^{\frac{1}{2}} .$$

From elementary geometrical considerations, we obtain the hydraulic radius of the rill as

$$r = \frac{1}{2} h \cos \varphi ,$$

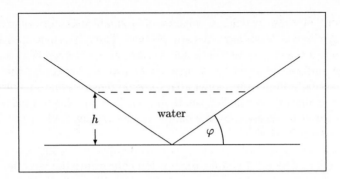

Fig. 1. Cross section through a v–shaped rill running downslope.

and thus the reference velocity given by the original Manning formula is:

$$v_r = \frac{1}{n} \left(\frac{1}{2} \cos \varphi \right)^{\frac{2}{3}} h^{\frac{2}{3}} \Delta^{\frac{1}{2}} .$$

Thus, our approach reproduces the correct law concerning roughness, water level, and slope even within a rill, but it overestimates the average velocity. If the walls of the rill are steep ($\varphi \approx 30°$), the overestimation is about 30 %; that is the price we have to pay for avoiding an explicit rill detection and using different approaches for sheet and rill flow.

The Differential Equation for the Surface Runoff

The conservation of mass results in the following differential equation for δ:

$$\frac{\partial \delta}{\partial t} + \operatorname{div}(\delta \, \boldsymbol{v}) = R - I \ . \tag{4}$$

R is the rainfall rate and I the infiltration rate. We do not take the evaporation rate into account here.

After putting the velocity $\boldsymbol{v}$ from equation (3) into equation (4) we obtain a parabolic nonlinear partial differential equation of second order for the thickness δ of the water film:

$$\frac{\partial \delta}{\partial t} = \operatorname{div} \left(\frac{1}{n} \, \delta^{\frac{5}{3}} \, |\boldsymbol{\nabla}(H + \delta)|^{\frac{1}{2}} \, \frac{\boldsymbol{\nabla}(H + \delta)}{|\boldsymbol{\nabla}(H + \delta)|} \right) + R - I \ . \tag{5}$$

Strictly speaking this differential equation is valid only in case of small slope inclinations (below 30 percent). It can be solved by using a fixed point method: for this iteration we treat the factor $\delta^{\frac{5}{3}}$ implicitly, whereas the δ in the gradient comes from the previous iteration step. Thus we obtain a nonlinear differential equation of first order in each iteration step. The fixed point method should converge to the exact solution of equation (5) linearly. If the water is, however, collecting within a hollow (or another uneveness), the inclination of the water surface becomes very small there. In this case a small change in the inclination of the water surface causes a large change in the flow velocity because the root function is not Lipschitz–continous in this region. Therefore convergence problems may occur. We take remedial action by changing the root function $r = |\nabla(H + \delta)|^{\frac{1}{2}}$ as follows:

$$r = \begin{cases} \sqrt{\Delta} & \text{for} \quad \Delta \geq A \\ \frac{\Delta}{\sqrt{A}} & \text{for} \quad \Delta < A \end{cases} \tag{6}$$

$$\text{with:} \qquad A = \text{real number}$$
$$\Delta = |\nabla(H + \delta)|$$

The maximum slope of this function is $1/\sqrt{A}$, and the fixed point iteration will converge if the following condition is met:

$$\Delta t \frac{\delta^{\frac{5}{3}}}{\sqrt{A}} < C \tag{7}$$

where the constant C depends on the spatial discretization. In order to avoid the problem that the maximum time step length Δt, where the iteration converges, decreases with increasing δ, the value A has to be chosen dependently on δ:

$$A \propto \delta^{\frac{10}{3}} \tag{8}$$

In order to avoid too small time steps as well as unrealistic behaviour within hollows, we choose the following function:

$$A = \left(\frac{\delta}{10 \text{ cm}}\right)^{\frac{10}{3}} \tag{9}$$

Fig. 2 shows the course of the new root function r for $\delta = 2.5$ cm and $\delta = 5$ cm. In case of a sheet flow the thickness δ of the draining water film rarely exceeds some millimeters. In this case the root function varies hardly because the value A becomes very small then. If the water is, however, collecting within a hollow, the flow depth δ can increase to several centimeters. In this case the root function changes for small slopes according to the above approach, so that no numerical instabilities occur any more. However a small artificial inclination of the water surface arises; but this inclination is hardly discernible and can be neglected without any problems.

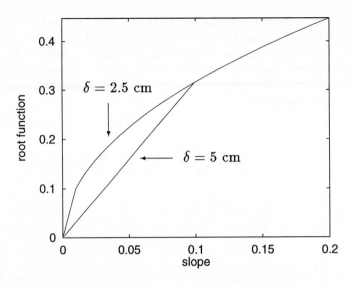

Fig. 2. Root function for $\delta = 2.5$ cm and $\delta = 5$ cm.

The Infiltration Model

In most soil erosion models, the infiltration process is based on a formulation of Green and Ampt (1911) that starts out from a uniform initial water content in the soil. The water finds its way into the soil along a near-horizontal front and fills up the pore volume immediately. The lateral flows are neglegted. For our small scaled runoff model however we want to decribe the infiltration process more precisely, additionally taking the lateral flows into account. The Richards equation (Richards 1931) is suitable for meeting these requirements. It results from a generalization of Darcy's law and the conservation of mass:

$$\frac{\partial}{\partial t}\left(\eta\,S(p)\right) + \sum_{i=1}^{2}\underbrace{\frac{\partial}{\partial x_i}\left(-K(p)\,\frac{\partial p}{\partial x_i}\right)}_{\text{lateral flows}} + \underbrace{\frac{\partial}{\partial x_3}\left(-K(p)\,\frac{\partial}{\partial x_3}(p + x_3)\right)}_{\text{vertical flow}} = 0 \quad (10)$$

with:

$$\eta = \text{porosity,}$$
$$p = \text{(standardized) pressure,}$$
$$S(p) = \text{saturation, and}$$
$$K(p) = \text{hydraulic conductivity.}$$

The fully implicit solution of this nonlinear three-dimensional equation requires a high numerical effort, especially if it shall be coupled with the surface flow equation (section 6). This effort can be reduced according to the following consideration: in the lateral direction the water spreads diffusively, whereas in the vertical direction the water flow is caused by the gravity additionally. That is why the lateral currents are quite small compared to the vertical currents and why there is no sharp saturation front in the lateral direction. However sharp saturation fronts may occur in the vertical direction. Thus, the pressure in the vertical currents shall be discretized implicitly, whereas the pressure in the lateral currents may be discretized explicitly. In this way the three–dimensional equation (10) can be reduced to a set of one-dimensional equations which can be treated independently from each other in each time step. Each nonlinear one-dimensional equation can be solved by using the newton method.

The soil is described by a finite volume grid that is refined and coarsened adaptively: in a zone around the saturation front the grid is refined. In the regions however where the saturation varies hardly the grid can be coarsened. This improves the convergence behaviour of the newton method significantly.

Coupling of Surface Runoff and Infiltration

In order to calculate the surface runoff according to equation (5) we need the infiltration rate I that is given by:

$$I = \left[K(p) \left(\frac{\partial p}{\partial x_3} + 1 \right) \right]_{surface} \tag{11}$$

The coupling of the runoff and infiltration model depends on the water amount on the relief surface. Two cases have to be considered:

1. The surface is water covered. In this case the soil water pressure at the upper edge of the slope must coincide with the hydrostatic pressure of the draining water film. Thus, the coupling is realized by using a Dirichlet boundary condition for the soil water flow: $p|_{surface} = \delta$.
2. The surface is dry or only moistened. There is no water film on the surface. Then the infiltration current corresponds to the maximum available water amount (according to the precipitation for example). Thus, the coupling is realized by using a Neumann boundary condition:

$$\left[K(p) \left(\frac{\partial p}{\partial x_3} + 1 \right) \right]_{surface} = \mathrm{div} \left(\frac{1}{n} \, \delta^{\frac{5}{3}} \, |\nabla(H+\delta)|^{\frac{1}{2}} \, \frac{\nabla(H+\delta)}{|\nabla(H+\delta)|} \right) + R \tag{12}$$

Some Examples

In the following examples we consider some artificial loamy slopes which are watered with a strong precipitation of some centimeters per hour. In this case the coefficient of roughness n is about $0.018 \, \mathrm{m}^{\frac{1}{3}} \mathrm{s}$. Each time we start out from a dry relief surface. Anyway the water in the unsaturated soil is in the hydrostatic equilibrium state first of all. The maximum hydraulic conductivity of the soil is about $1 \, \mathrm{cm/h}$. In a depth of several meters there is a layer that cannot be

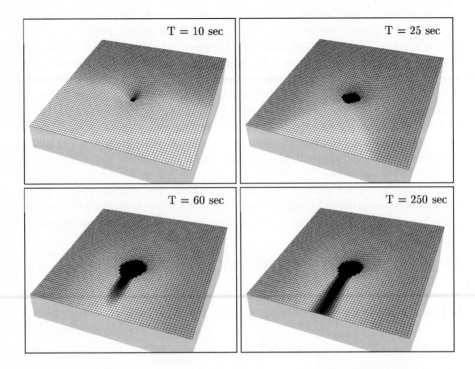

Fig. 3. Runoff through a hollow induced by an instantaneous inflow.

penetrated by water (for example a clay layer). At the foot of the respective slope the water can drain over the relief edge in a stream (for example). In order to distinguish the different thicknesses δ of the water film we use a greyscale for the grid cells. The different saturations in the soil are presented by a greyscale as well. The more water is present the darker the grey is chosen.

The first example shows that our model copes with an arbitrary uneveness like a hollow. We consider a slope with an inclination of 20 percent and a surface of about 36 m^2. In the middle of this slope is a hollow with a diameter of about 1 meter (Fig. 3). In order to illustrate the effect of the hollow we do not water the slope with a precipitation, but with a continous inflow of 1.6 l/s over the upper edge of the surface. Anyway, for the sake of simplicity, the infiltration rate is setted equal to zero in this example. The hydrograph presented in Fig. 4 shows the total outflow over the lower edge of the surface.

After 10 seconds the water reaches the hollow. The sharp water front that is caused by the continous inflow has been smoothed clearly. After 25 seconds the water arrives at the lower edge of the relief surface. This effects the increase in the hydrograph after 25 seconds that is delayed due to the smoothed water front. After 60 seconds the hollow starts overflowing, so that the hydrograph rises again. After 250 seconds the runoff is stationary, and the hydrograph has almost reached the inflow value of 1.6 l/s.

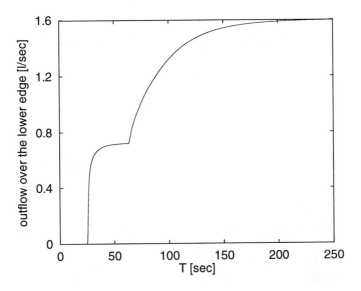

Fig. 4. Outflow over the lower edge versus time.

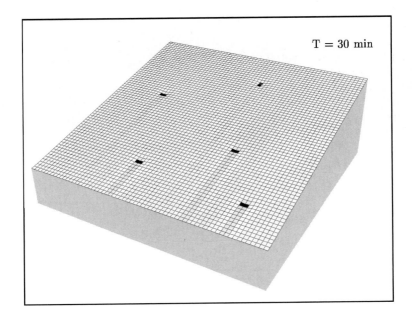

Fig. 5. Surface runoff on a slope with five weak hollows.

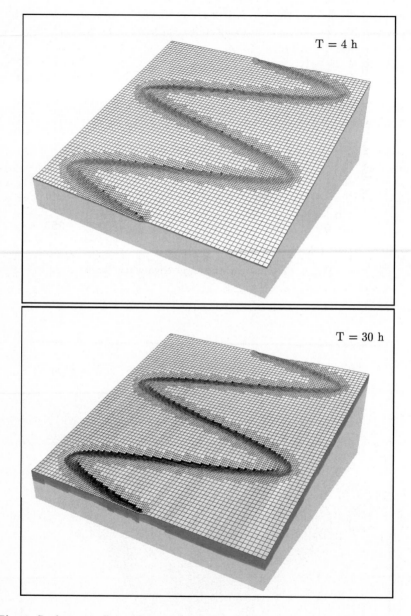

Fig. 6. Surface runoff on a loamy slope with a path consisting of loamy soil, too.

In the second example we consider a slope that covers up an area of about 25 m^2. Its inclination is 20 percent; the surface grid consists of 128 cells like in the first example. There are five weak hollows which are randomly distributed. Within these hollows the maximum hydraulic conductivity of the soil is reduced by about 40 percent. Thus one can interpret these hollows as prints of cows or cattles for example. The precipitation rate is about 5 cm/h. After 30 minutes some water has already collected in the hollows, and several hollows overflow (Fig. 5). This example shows the sensitivity of the surface runoff with regard to locally variing infiltration rates.

The third example shows a slope on which a path winds its way downwards. The surface is described by a grid that consists of about 4 000 cells and that covers up an area of about 600 m^2. We choose a precipitation of about 2 cm/h. After 4 hours the surface runoff starts on the path (Figure 6). About 30 hours later the runoff has almost stretched over the whole slope, and the saturation front is discernible clearly in a depth of about one meter in the soil (lower part of Fig. 6).

As the fourth example we consider the same slope, but with a tarred path. Thus the water cannot infiltrate on the path, and the surface runoff begins there considerably earlier as in the previous example, namely after 30 seconds. About 30 hours later the saturation front is discernible in the soil clearly (Figure 7). However the water cannot get into the soil evenly as in the previous example

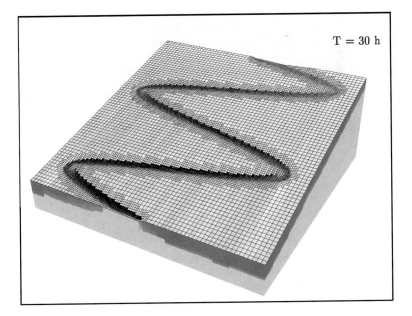

Fig. 7. Surface runoff on a loamy slope with a tarred path.

with the loamy path, because it can reach the region under the path by lateral flows only. This affects the shape of the saturation front close to the path.

Conclusions

In contrast to many other large scaled runoff models we have developed a model that enables us to investigate the influence of small scaled surface structures like hollows and rills on the runoff. Such an investigation is important because, even in the case of a large scaled consideration, the runoff and the soil erosion can be affected by such small scaled structures significantly. The infiltration process has to be described in our model more precisely than in other models like Green and Ampt (1911) because it plays a dominant part as well with respect to the runoff.

Acknowledgment. The authors would like to thank the German Research Foundation (DFG) for financial support within the Collaborative Research Center 350.

References

Abrahams, A.D., A.J. Parsons, and J. Wainwright (1994): Resistance to overland flow on semiarid grassland and shrubland hillslopes. Journal of Hydrology, 156:431–446.

Abrahams, A.D., G. Li, and A.J. Parsons (1996): Rill hydraulics on a semiarid hillslope, Southern Arizona. Earth Surface Processes and Landforms, 21:35–47.

de Roo, A.P.J., C.G. Wesseling, N.H.D.T. Cremers, R.J.E. Offermans, C.J. Ritsema, and K. van Oostindie (1994): LISEM: a new physically–based hydrological and soil erosion model in a GIS–environment, theory and implementation. In: Olive, L.J., R.J. Loughran, and J.A. Kesby: Variability in stream erosion and sediment transport, Int Assoc. Hydrol. Sci. Publ. 224, Wallingford.

Emmett, W.W. (1970): The hydraulics of overland flow on hillslopes. U.S. Geological Survey Paper, A:662.

Giles, R. (1962): Theory and problems of fluid mechanics and hydraulics. McGraw–Hill Inc..

Govers, G. (1992): Relationship between discharge, velocity, and flow area for rills eroding in loose, non–layered material. Earth Surface Processes and Landforms, 17:515–528.

Govindaraju, R.S., S.E. Jones, and M.L. Kavvas (1988): On the diffusion wave modeling for overland flow, 1, solution for steep slopes. Water Resour. Res., 24(5): 734–744.

Green, W.H., and G.A. Ampt (1911): Studies on Soil Physics: 1. Flow of air and water through soils. Journal Agric. Sci., 4:1–24.

Hergarten, S., and H.J. Neugebauer (1997): Homogenization of Manning's formula for modeling surface runoff. Geophys. Res. Lett., 24(8):877–880.

Knisel, W.G. (1980): CREAMS – a field scale model for chemicals, runoff, and erosion from agricultural management systems. USDA Conserv Res Report 26, US Dept of Agriculture/Science and Education Administration, Washington, D.C.

Lane, L.J., and M.A. Nearing (1989): USDA – water erosion prediction project: hillslope profile model documentation. NSERL Report 2 (USDA-ARS National Soil Erosion Laboratory), West Lafayette.

Morgan, R.P.C. (1989): USDA – water erosion prediction project: hillslope profile model documentation. NSERL Report 2 (USDA–ARS National Soil Erosion Laboratory), West Lafayette.

Morris, E.M., and D.A. Woolhiser (1980): Unsteady one–dimensional flow over a plane: partial equilibrium and recession hydrographs. Water Resour. Res., 16(2): 355–360.

Pearce, A.J. (1976): Magnitude and frequency of erosion by Hortonian overland flow. Journal of Geology 84:65–80.

Richards, L.A. (1931): Capillary conduction of liquids in porous mediums. Physics, 1:318–333.

Schmidt, J. (1991): A mathematical model to simulate rainfall erosion. In: Bork, H.R., J. de Ploey, and A.P. Schick: Erosion, transport and deposition processes – theory and models.

Smith, R.E. (1992): OPUS – an integrated simulation model for transport of nonpoint–source pollutants at the field scale. I: documentation, US Dept of Agriculture – Soil Conserv Service 98, Washington, D.C.

von Werner, M. (1996): GIS–orientierte Methoden der digitalen Reliefanalyse zur Modellierung von Bodenerosion in kleinen Einzugsgebieten. Dissertation, FU Berlin.

Weltz, M.A., A.B. Arslan, and L.J. Lane (1992): Hydraulic roughness coefficients for native rangelands. Journal of Irrigation and Drainage Engineering, 118:776–790.

Woolhiser, D.A., R.E. Smith, and D.C. Goodrich (1990): KINEROS, a kinematic runoff and erosion model: documentation and user manual. USDA–ARS–77, US Dept of Agriculture – Agriculture Res Service, Washington, D.C.

Zhang, W., and T.W. Cundy (1989): Modeling of two–dimensional overland flow. Water Resour. Res., 25(9): 2019–2035.

A Dupuit Approximation for Saturated–Unsaturated Lateral Soil Water Flow

C. Blendinger

Institute of Applied Mathematics, University of Bonn, Germany

Abstract. A generalized Boussinesq equation is derived to describe saturated–unsaturated lateral flow in porous media where the wellknown Dupuit approximation is not valid because of a relatively large capillary fringe or unsaturated zone. It is based on a singular perturbation of the porous media equation (Darcy law) on a thin domain. Infiltration and ponding are also included. The resulting approximation is determined by a lowerdimensional equation with some numerical advantages. Proofs are available. These help to overcome some of the problems appearing with empirical or overparametrized models.

In hydrogeological literature may be found many mathematical models and computer programs to simulate the hydrological cycle on a hillslope or catchment scale (see examples below). They try to incorporate the phenomena of infiltration of precipitation, saturated–unsaturated flow in an inhomogenous soil, interaction with groundwater and drainage channels, surface runoff, channel flow, evapotranspiration and sometimes erosional effects. Often there are more or less physically based mathematical models for these different phenomena on a local scale. Constructing global models on the regional scale one has to be aware of two main problems: First there is need to integrate those local models to a consistent description on the global scale, which means in praxi unique solvability or stable computability. A model consisting simply of an accumulation of the detailled models of the different local processes would include a huge amount of parameters. Hence another main purpose constructing a global model is parameter reduction. There are several reasons to do this process sometimes called upscaling: At first there is in any case not enough measured data available to get reasonable values for all these parameters or retrieving all that data would be too expensive. Secondly because such overparametrized models are very insensitive on the change of most of the incorporated parameters, they do not provide much understanding on the governing rules on the larger scale and therefore have restriced reliability in forecast. A widely used way out is choosing empirical relations on the larger scale, whose connection to the locally assumed laws is often not mathematically justified. Another reason, which should have lost importance in the last decade, is the lack of computer power and sometimes numerical techniques to get numerical solutions on large domains and over long time intervalls of the partial differential equations governing a physically based model. But even because modern computers allow the tackling of models with many degrees of freedom, extra attention on the used equations and their theoretical justification is indispensible. In the following it will be shown for the processes of saturated–unsaturated flow and infiltration, that application of mathematics will help to justify the upscaling and integrating different parts of the hydrolgical cycle into a reasonable approximation with reduced computational effort.

1 Some Models for Saturated–Unsaturated Flow

Looking a bit closer to some models of flow in the saturated–unsaturated zone, the ways of integrating local information and scaling up the rules are very different between models.

A first attempt to overcome the need to solve the full three–dimensional equations was presented by Pikul et al. (1974) and was later implemented in the SHE model (Abbott et al. 1986). It is based on a decoupling of one–dimensional vertical unsaturated flow (Richards equation) from two–dimensional lateral saturated flow (Boussinesq equation). The interaction between the two parts is such that the groundwater table serves as lower boundary for the vertical flow and input from the unsaturated zone changes the free grondwater surface as a source term in the Boussinesq equation. From a mathematical point of view it is not quite clear that such a coupling defines a wellposed problem, whose numerical solutions are of physical relevance. On the other hand is the use of the Boussinesq equation for lateral saturated flow restricted to regimes where the vertical extension of the saturated zone is large compared to capillary fringe and unsaturated zone as noted already by Vachaud and Vauclin (1975) (see also the reply of Pikul et al. 1975). The case of temporal vanishing of the saturated zone (which may well happen in interflow–like regimes) is also not handled within the usual Boussinesq equation.

A widely used program is TOPMODEL which is based on ideas in Beven and Kirkby (1976). It describes the lateral flow as a combination of the continuity equation with an empirical relation between flow and saturation deficit, which leads after some restrictive assumptions and the introduction of some empirical parameters to a more or less explicit formula for lateral flow. This ansatz is based on a model for strip–like hillsope sections and has no natural generalisation to catchment–like geometries. Instead the catchment is modeled as an assemblage of such hillslope sections. Even if the flow computed with this formula is very well reproduced, the recomputation of the local seepage is delicate and not of much a priori accuracy. But despite its few parameters it may be fitted very well on many realistic data as found in the literature compiled on the TOPMODEL homepage[1]. Therefore it is an example of choosing useful empirical relations on the larger scale, whose foundation on the locally valid physical laws and conditions of validity are nevertheless not very well established in terms of mathematical analysis.

Another computationally simple approach is used in cascadic storage and routing models like HSPF (Johanson et al. 1984) or PRMS (Leavesley et al.

---

[1] http://es–sv1.lancs.ac.uk/es/FREEWARE/FREEWARE.html

1983). They consist of more or less empirical relations with many parameters whose physical meaning are often poorly understood. Because of the discrete nature of these models there is also a strong dependence of the parameters on the chosen grid. The concept of Hydrological Response Units (Flügel 1995) is an attempt to the vast amount of parameters in such overparametrized models, but this does not touch the other criticisms.

2   Just Another Model – but with Proof

A new approximation for saturated–unsaturated flow is presented. Its basic scaling idea is due to Luckhaus (1993), but all detailed (and herein omitted) mathematical analysis is found in Blendinger (1996). It tries to overcome many of the stated difficulties: it is physically based, because it starts with the experimentally and partly mathematically established Darcy law as the constitutive relation between pressure gradient, pressure and flux. Its central ingredient is a partial differential equation which may be solved with stable and established numerical methods. It uses a provable property of the vertical pressure profile such that in simple scenarios no extra equation need to be solved for vertical unsaturated flow. Therefore it includes only lateral space variables as in the Boussinseq equation. It provides a uniform approch to two–dimensional hillslope sections and three–dimensional catchment areas. It handles the limit cases of vanishing saturated zone and reaching saturation up to top of the soil and remains valid if the saturated zone is small compared to capillary fringe and unsaturated zone. Although the new effective parameters are in general not computable from local data (because of lack of these data), the model provides some insight to the structure of the constitutive law which should be used on the large scale. The validity of this law is mathematically proved and not only empirically assumed.

It should be noted here that two important features of the problem are ignored in the mathematical analysis: strong local heterogeneities in soil parameter functions, which should be 'scaled away' into some effective (and hopefully simple structured) parameters and the influence of evapotranspiration, which keeps the model valid as stated only for time periods without vegetational growth. The first is excluded for mathematical reasons because there are only some assumptions but no strict proof on the exact structure of the limiting law in the nonlinear case even in geometrically simpler cases than considered here. Evapotranspiration as a sink term is ignored mainly because it is not quite clear how the known models for this term (e.g. Penman–Monteith, see Maidment 1992) would fit in the scaling used to derive the model. Another completely ignored phenomenon is hysteresis.

The main indispensible geometrical assumption is the existence of an impermeable layer (e.g. bedrock or loam) defining a zero flux boundary condition at the lower boundary of the soil profile. Another geometric feature of the applications in mind is that lateral extension of the flow domain is much larger than the vertical lengthscale measured as distance between soil surface to impervious layer. This is valid for areas like the Sieg catchment (Germany, north–east of Bonn), situated in a typical middle mountain range, where over an area of approx. 2800 km^2 the soil profiles are not thicker than 1 m (P. Dornberg, SFB 350, University of Bonn, personal communication). The apparent elevations are between 50 m and 500 m; hence one may think of the hillslopes as 'small'.

To illustrate the ideas stated later in formulas in Fig. 1 two sketches of a numerical example of a prototypic two–dimensional situation are shown. They are computed with the program silVlow (Blendinger 1995) and visualized with help of the graphics package GRAPE^2. It consists of a hillslope section with zero flux boundary condition at the lower boundary, temporal rainfall infiltration at the upper boundary, a water divide at the left lateral boundary and a time–independent Dirichlet condition at the right lateral boundary, which approximates the interaction with a free water reservoir or river. The pictures show flux vectors (arrows, whose lengths are proportional to flux rates). The scaling of the arrows differs (with a factor of about 5) between the upper and the lower picture for reasons of displayability. The solid lines in the two pictures are isolines of the total potential, on which the flux vectors are perpendicular. The flow is driven from the gradient of this total potential and the density of these isolines is a measure of the amount of flow. The scaling of the isolines is the same on both pictures. The upper picture shows a typical flow profile at a time far away from the last rainfall event: the total potential is almost completly determined by the geometry of the lower (impervious) boundary, which implies parallelity of flux vectors to that boundary. What is happening during a (short) rainfall event, is show in the lower picture: rapid vertical flow in the upper (unsaturated) zone, whereas the slow lateral flow in the saturated lower part is not much affected. If the rainfall events happen seldom relative to the relaxation time of the vertical flow, most of the time the flow regime is like in the upper picture and an equation which integrates over the vertical coordinate seems to be a valid approximation. The short term rainfall events then serve as a source to keep mass balance.

3   From Darcy's Law to a Dupuit Approximation

These short hydrogeological considerations may be reformulated in a mathematically tractable problem: *Give a reasonable approximation of saturated–unsaturated flow in thin domains with moderate slope and episodic short term rainfall events.*

The technique to derive the proposed approximation is well known as 'singular perturbation' (see e.g. Eckhaus 1979). It consists of three steps, which read in the present context as follows:

1. Use known equations for flow in the given domain.
2. Embed the original problem in a scale of problems depending on a small parameter $\epsilon > 0$.
3. Take the limiting problem for $\epsilon \to 0$ as an approximation to the original problem, which is given as one of the scaled problems for small ϵ.

---

^{2} developed at SFB 256, University of Bonn

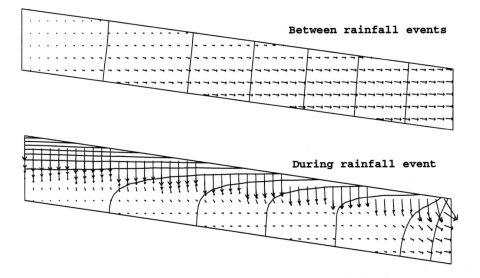

Fig. 1. Numerical example: Isolines of total potential and flux vectors (Arrows not on the same scale, see text)

The 'singularity' of the perturbation here is that the limiting equation will be defined on a domain with smaller dimension than the original equation. The three steps shall be explained now in some detail.

3.1   Saturated–Unsaturated Flow and Infiltration

The well known porous media equation is used to describe the saturated–unsaturated flow in the given domain (for a detailled derivation see Bear 1979 or Collins 1961). It is a combination of the continuity equation with the Darcy law (for some theoretical justification of this historically empirical found relation see Dagan 1989 or Hornung 1992) and is written here as

$$\partial_{\bar{t}}\theta(p(x,z,\bar{t})) - \nabla \cdot K(\theta(p(x,z,\bar{t})))\nabla(p(x,z,\bar{t}) - z) = 0 \; , \tag{1}$$

where the independent variables are $\bar{t}$ for time, x for the lateral space variables and z for the vertical space variable (pointing downward). The dependent variable p is hydrostatic pressure in the saturated zone and matrix potential in the unsaturated zone measured in the unit of a length (height of water column). Ignoring osmotic and other effects the total potential consists only of pressure and gravitation and is therefore $p - z$. The function $\theta(p)$ describes volumetric water content and $K(\theta)$ is a conductivity function, both (partly for simplicity) homogenous and isotropic in space.

For this nonlinear degenerate parabolic–elliptic equation one needs an initial condition for θ and boundary conditions on the whole bondary to get a wellposed

problem (see Alt and Luckhaus 1983). At the impervious layer and at water
divides locateted at lateral boundaries a zero flux (Neumann) condition

$$j \cdot \nu = 0$$

is natural, where $j := -K(\theta(p))\nabla(p-z)$ is the flux vector and hence $j \cdot \nu$ denotes
the normal flux in direction of the outer normal ν. At the part of the boundary
where interaction with a reservoir is assumed, as a first choice (see Bear 1979)
a pressure (Dirichlet) condition

$$p = p_D$$

for some given function p_D is posed. A more realistic choice would be a Signiorini
condition as in Alt et al. (1984). If one takes at the infiltration boundary a simple
nonhomogenous Neumann condition

$$j \cdot \nu = -q/\gamma \ , \tag{2}$$

where q is rainfall intensity and γ the local area element of the surface (a purely
geometrical correction term for the rainfall measured on a horizontal reference
plane), one runs into theoretical and even numerical trouble, because at very
high rainfall rates or if saturation reaches top of the soil, this would produce
unphysically steep pressure gradients. The way out is to model ponding and surface runoff explicitly as Filo and Luckhaus (1997) did. Here only a simplification
of this ansatz is used. It neglegts lateral surface runoff and deals with ponding
only. This is a reasonable approximation if the slope of the surface is small or
the height of the ponding water is on the scale of the natural surface roughness,
which is already smoothed out in the used descripion of the surface (see Filo and
Luckhaus 1995 for some mathematical justification of this smoothing). Hence let
w denote the height of ponding at every point of the infiltration boundary. Then
the time behavior of w is goverened by the differential equation

$$\partial_{\tilde{t}} w = j \cdot \nu + q/\gamma \tag{3}$$

and the coupling between w and p at the surface is given as

$$w \geq 0, \quad p \leq 0, \quad w \cdot p = 0 \ . \tag{4}$$

This states that there is either ponding $(w > 0)$, from which the Dirichlet condition $p = 0$ follows, or no ponding $(w = 0)$: if this is valid on some time
interval, one furthermore has $\partial_{\tilde{t}} w = 0$, hence from (3) the infiltration condition
(2) follows. In hydrological literature (e.g. Maidment 1992) this is a well known
but sometimes more operationally formulated infiltration model. It is proved in
Blendinger (1996) that these boundary conditions lead to a solvable problem.

3.2   The Scaling

The scale of problems is essentially defined as a scale of different geometries.
The small parameter ϵ will be the relation of vertical to lateral extension of the
domain. If vertical extension is of order 1, lateral extension will be of order $1/\epsilon$.

Let Q be some one– or two–dimensional domain and g_1, g_2 two sufficiently smooth functions parametrizing the lower resp. upper boundary of the reference domain Ω; in formula:

$$\Omega = \{(x,z) \in Q \times \mathbb{R} | g_2(x) \leq z \leq g_1(x)\} \ .$$

Then the scaled domains Ω^ϵ for $\epsilon > 0$ are given as

$$\Omega^\epsilon := \{(\bar{x}, z) \in \mathbb{R}^{n-1} \times \mathbb{R} | (\epsilon\bar{x}, z) \in \Omega\} \ .$$

For Ω beeing a parallelogram (and Q simply a one–dimensional interval) the related Ω^ϵ for $\epsilon = 0.5, 0.2, 0.1$ are shown in Fig. 2. This shows that the scaling produces long hillslopes with decreasing slope angles. It should be noted here, that the described approch by no means uses any dimension specific properties and keeps valid with the same formulas for a two–dimensional (or even higher dimensional) Q . On every Ω^ϵ the unscaled equation (1) is assumed.

Because of the chosen form of Ω^ϵ one may use the variable transform $x = \epsilon\bar{x}$ to derive an equivalent ϵ–dependent equation and accordingly transformed initial and boundary conditions on the fixed reference domain Ω. The porous media equation (1) transforms into

$$\partial_{\bar{t}}\theta(p) - \epsilon^2 \nabla \cdot K^\epsilon(\theta(p))\nabla(p-z) = 0 \ ; \tag{5}$$

because K was assumed to be scalar, the permeability of the transformed equation is

$$K^\epsilon(\theta(p)) = \begin{pmatrix} K(\theta(p))I_{n-1} & 0 \\ 0 & \frac{1}{\epsilon^2}K(\theta(p)) \end{pmatrix} \ ,$$

where I_{n-1} denotes the $(n-1) \times (n-1)$-unit matrix. Because the formal limit of (5) does not provide any information on lateral flow, the time variable $\bar{t}$ in (5) needs to be transformed as $t = \epsilon^2\bar{t}$. This time scaling is motivated from the fact that the length of the hill (the water has to flow down) scales with $1/\epsilon$ and the driving force, the slope of the lower impermeable boundary (determining the speed of flow), also scales with $1/\epsilon$. A precise justification is the convergence proof in Blendinger (1996), because it shows that this scaling leads to a reasonable approximation. The central equation scaled in space and time is therefore

$$\epsilon^2\partial_t\theta(p^\epsilon) - \epsilon^2 \nabla \cdot K^\epsilon(\theta(p^\epsilon))\nabla(p^\epsilon - z) = 0 \ . \tag{6}$$

Here the ϵ-dependence is explicitly stated by the denoting the solution p^ϵ. If the ϵ–problem is solved until time T, the unscaled problem is solved until time $\bar{T} = T/\epsilon^2$.

3.3   The Limit Equation

Now a series expansion $p^\epsilon = p_0 + \epsilon^\alpha p_1 + \ldots$ with some $\alpha > 0$ is used with the hope to take p_0 as good approximation for p^ϵ even for small ϵ. As p_0 is the leading term, one gets an equation for p_0 as the formal limit $\epsilon \to 0$ in (6):

$$-\partial_z K(\theta(p_0))\partial_z(p_0 - z) = 0 \ , \tag{7}$$

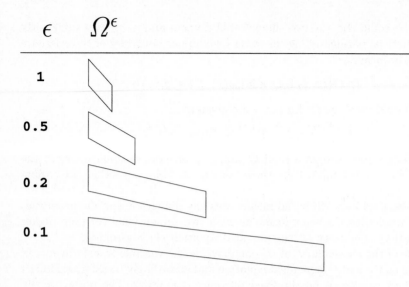

$\epsilon \qquad \Omega^\epsilon$

1

0.5

0.2

0.1

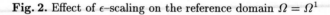

Fig. 2. Effect of ϵ–scaling on the reference domain $\Omega = \Omega^1$

from which follows after integration in z-direction

$$-K(\theta(p_0(x,z,t)))\partial_z(p_0(x,z,t)-z)|_{z=g_2(x)}+K(\theta(p_0(x,z,t)))\partial_z(p_0(x,z,t)-z)=0$$

for every z between $g_1(x)$ and $g_2(x)$; here t and x are parameters. For times t, at which in the limit case at the infiltration boundary $\{z = g_2(x)\}$ the condition

$$-K(\theta(p_0(x,z,t)))\partial_z(p_0(x,z,t) - z)|_{z=g_2(x)} = 0$$

is fulfilled (which may be rewritten as a zero flux condition), one gets because of $K > 0$ the equation

$$\partial_z(p_0(x,z,t) - z) = 0 \tag{8}$$

for all (x, z) in the domain Ω. Hence p_0 is linear in z and therefore determined, if the value $p_0(x, g_1(x), t)$ at the impervious bondary is known.

The equation for lateral flow in the transformed t–time follows formally from (6) after division by ϵ^2 and with the additional assumption

$$\frac{1}{\epsilon^2}\partial_z(K(\theta(p^\epsilon))\partial_z(p^\epsilon - z)) \to 0 \text{ for } \epsilon \to 0 \; ;$$

the seeked timedependent equation in Ω is

$$\partial_t\theta(p_0) - \nabla_x \cdot K(\theta(p_0))\nabla_x(p_0 - z) = 0 \; . \tag{9}$$

Here the subscript at the gradient operator ∇_x denotes that the partial derivatives are taken only with respect to the x–variables. Equation (9) for p_0 may be integrated over z because (8) describes the behavior of p_0 in z–direction. This gives for the desired $H(x,t) = p_0(x, g_1(x), t)$, the pressure at the lower boundary, in Q the equation

$$\partial_t \tilde{\theta}(H, x) - \nabla_x \cdot \tilde{K}(H, x) \nabla_x (H - g_1) = q \ . \tag{10}$$

The source term q in (10) is simply the rainfall intensity used in (3) if the rainfall intensities for the ϵ–problems are assumed to be scaled with ϵ^2. The solution p_0 of (9) is because of (8) related to H via

$$p_0(x, z, t) = H(x, t) + z - g_1(x) \tag{11}$$

and the new parameter functions $\tilde{\theta}, \tilde{K}$ are computed from θ, K^x as

$$\tilde{\theta}(H, x) := \int_{H+g_2(x)-g_1(x)}^{H} \theta(p) \mathrm{d}p$$

resp. as

$$\tilde{K}(H, x) := \int_{H+g_2(x)-g_1(x)}^{H} K^x(\theta(p)) \mathrm{d}p \ .$$

The space dependency of $\tilde{\theta}$ resp. $\tilde{K}$ results from the local variation of the soil thickness $g_1 - g_2$. The formulas keep valid if there is some weak spacial heterogeneity of θ or K^x, which means no oszillation of order ϵ in Ω. This is violated in the case of interest, where the heterogeneity is of order 1 in Ω^ϵ.

3.4 The Infiltration Boundary Condition

Up to here the previous considerations are focused on the limit of the partial differential equation (6) and ignore almost completely the boundary conditions. The detailled analysis shows that the ponding conditions (4) persist in the limit as follows: There is an upper limit for H of the form $H_{\max} := g_1 - g_2$, which is expected in light of (11). The height of ponding is described as a function W defined on Q. The conditions (4) transform into

$$H \leq H_{\max}, W \geq 0, W \cdot (H - H_{\max}) = 0 \ . \tag{12}$$

As long as no ponding appears, there is $W = 0$ and (10) governs the behavior of H. If water begins to pond, H admits the value $H_{\max}$ and the evolution of W is determined by the ordinary differential equation with x as a parameter

$$\gamma(x) \partial_t W(x, t) = q_0(x, t) - \nabla_x \cdot (\tilde{K}(H_{\max}(x)) \nabla_x g_2(x)) \ , \tag{13}$$

which contains the curvature of the infiltration boundary $\{z = g_2(x)\}$ in the rightmost term. As in (2) the term $\gamma(x)$ denotes the area element of the upper boundary. The lateral Dirichlet and Neumann conditions persist as such.

3.5   The Dupuit Approximation for Saturated Flow

The usual Boussinesq equation may be formally derived from (10), if θ degenerates torwards a Heavyside function and K admits only two values: for some fixed $K_s, \theta_s > 0$ let θ, K given as

$$\theta(p) := \begin{cases} \theta_s & \text{if } p \geq 0, \\ 0 & \text{else} \end{cases} \quad , \quad K(\theta) := \begin{cases} K_s & \text{if } \theta = \theta_s, \\ 0 & \text{else} \end{cases} .$$

This ignores flow in the capillary fringe and in the unsaturated zone. Then the new parameter functions are linear for $H \geq 0$:

$$\tilde{\theta}(H) = \theta_s H, \quad \tilde{K}(H) = K_s H .$$

Hence (10) transforms into Boussinesq equation

$$\theta_s \partial_t H - \nabla_x \cdot K_s H \nabla_x (H - g_1) = q . \tag{14}$$

A rigorous analysis of this approximation of porous media flow by a free boundary problem is made in Alt et al. (1984).

Another way to derive the Boussinesq equation is explaind in Bear (1979). He starts from saturated flow (linear Darcy law) and gets (14) from the Dupuit assumption concerning the direction of flux vectors. Because (7) is a generalization of this assumption in a

saturated–unsaturated zone, it is justified to call the solution of (10) a Dupuit approximation. It should be remarked that the mathematical theory in Blendinger (1996) shows that (7) is in fact a consequence of the chosen ϵ–scaling and needs not to be assumed.

4   Conclusion

All formal limits above are rigorously proved in Blendinger (1996) using a weak formulation of the partial differential equation. The main result is the solvability of the obstacle problem (10), (12), (13) with an initial condition and Dirichlet–Neumann boundary conditions. Most of the needed a priori estimates use essentially the properties of the ϵ–scaling. Other mathematical tools are time discretisation, regularization of free boundary problems, a barrier construction, monotone operators and a compensated compactness argument.

As already mentioned one of the main properties of the proposed approximation is dimension reduction: instead of the full three–dimensional partial differential equation (1) only the two–dimensional equation (10) needs to be solved in catchment–like situations. In hillslope–like geometries the reduction is from a two–dimensional to a one–dimensional equation. In the computed examples the one–dimensional approximating equation is solved about 20 times faster than the full two–dimensional equation. The gained computer time may be used e.g. for extra computations concerning parameter fitting, long time behavior, heterogeneity or testing scenarios. In case of no ponding the partial differential equation

(10) is because of the structure of the parameter function $\tilde{\theta}$ strictly parabolic and on contrary to (1) no elliptic degeneration appears. Hence the resulting discrete equations are much better conditionend and therefore fast solvable with iterative algorithms like the conjugate gradient method. In case of ponding no spacial coupling appears and W may be computed independently point by point. These two cases need not implemented with case distinctions if numerical regularizing techniques for variational inequalites (see Glowinski 1984) are used.

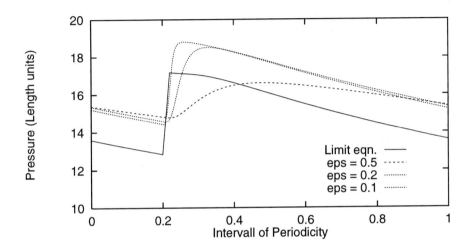

Fig. 3. Numerical example: Pressure at lower boundary (impervious layer)

The outstanding problems with strong heterogeneity and evapotranspiration might be examined with numerical experiments before trying to prove strict results. From a mathematical point of view the questions of uniqueness and error estimates or convergence rates remain open, but a short impression of the quantitative behavior of the Dupuit approximation is given in Fig. 3, Fig. 4 and Fig. 5. It shows some numerical results of the full two–dimensional equation (1) for ϵ and Ω^{ϵ} as shown in Fig. 2 computed with silVlow (Blendinger 1995) and the related one–dimensional Dupuit approximation computed with a simple finite element code. Short term rainfall is chosen periodically and the pictures show a period after relaxation of the inital condition (the data used for Fig. 1 are also taken from these computations for $\epsilon = 0.1$).

Figure 3 displays the pressure at the point in the mid of the lower (impervious) boundary of the domain. Here and in the figures below the interval of periodicity ranges from 0 to 1. The pressure scale is a lengthscale – think of cm water column. Because the chosen geometry has a thickness of 100 units, a pressure value of say 20 means that a fifth of the vertical profile is saturated. Despite some systematic deviation the time behavior is increasingly well captured if ϵ gets smaller.

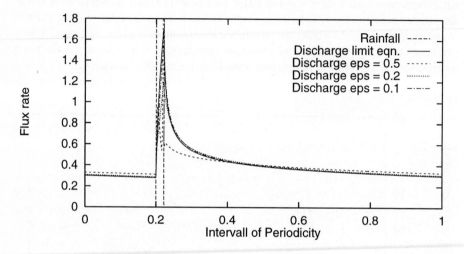

Fig. 4. Numerical example: Input and output flux rates, whole intervall of periodicity

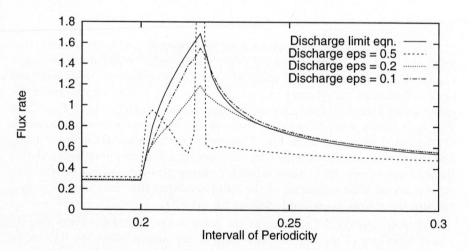

Fig. 5. Numerical example: Input and output flux rates, detail of Fig. 4

Figure 4 shows the rainfall and the discharge at the lateral Dirichlet boundary over the whole interval. On the chosen scale the rainfall event admits a flux rate of 20 over the relatively short interval denoted by the two vertical dotted lines; this value is not displayable in the figures. In Fig. 5 the relevant part of the interval around the rainfall event is shown in detail.

It is apparent that the approximating equation overestimates discharge at rainfall. Therefore the storage is underestimated by the approximation, which is consistent with the systematic deviation of the pressure data in Fig. 3. It is remarkable that the fit of the discharge is quite well at most of the recession even for relatively large ϵ as 0.2 or 0.1. The somewhat surprising behavior for the ϵ-value of 0.5 is mainly due to boundary effects at the lateral Dirichlet boundary and proves that this case should not approximated with the asymptotic limit equation.

But even without this numerical evidence for usefulness of the proposed model much is gained because of the available proofs. They show conditions and restrictions of validity of the model a priori. Therefore the main reason using experimental data is not fitting the parameters but looking at the assumptions the model and the measurements are based on. This will support progress in identifying the relevant parameters on the different scales and their relations.

Acknowledgments. I would like to thank the Collaborative Research Center (SFB) 350 for the opportunity to develop and present this material. The participants of the Workshop and their fruitful hints and contributions in discussion, esp. from M. J. Kirkby and J. Filo, helped me a lot to improve this written version.

References

Abbott, M.B., J.C. Bathhurst, J.A. Cunge, P.E. O'Connell, and J. Rasmussen (1986): An Introduction to the European Hydrological System – Systeme Hydrologique Europeen, 'SHE', 2: Structure of a physically–based, distributed modelling system. J. Hydrol. 87:61–77.

Alt, H.W., and S. Luckhaus (1983): Quasilinear elliptic–parabolic differential equations. Math. Z. 183:311–341.

Alt, H.W., S. Luckhaus, and A. Visintin (1984): On nonstationary flow through porous media. Ann. Math. Pura Appl. 136:303–316.

Bear, J. (1979): Hydraulics of Groundwater. McGraw-Hill, New York.

Beven, K.J., and M.J. Kirkby (1976): Towards a simple physically–based variable contributing area model of catchment hydrology. Working Paper 154, School of Geography, University of Leeds.

Blendinger, C. (1995): silVlow – A program for the computation of saturated–unsaturated flow in twodimensional porous media (in German), Preprint no. 13, SFB 350, University of Bonn.

Blendinger, C. (1996): An approximation of saturated-unsaturated Darcy flow in thin domains (in German). Preprint no. 485, SFB 256, University of Bonn.

Collins, R.E. (1961): Flow of Fluids through Porous Material. Reinhold, New York.

Dagan, G. (1989): Flow and Transport in Porous Formations. Springer, Berlin.

Eckhaus, W. (1979): Asymtotic Analysis of Singular Perturbations. Studies in mathematics and its applications 9, North–Holland, Amsterdam.

Filo, J., and S. Luckhaus (1995): Asymptotic Expansion for a Periodic Boundary Condition. Journal of Differential Equations 120(1):133–173.

Filo, J., and S. Luckhaus (1997): Modelling of surface runoff and infiltration of rain by an elliptic-parabolic equation coupled with a first order equation on the boundary. Submitted.

Flügel, W.-A. (1995): Delineating Hydrological Response Units by Geographical Information System Analyses for Regional Hydrological Modelling Using PRMS/MMS in the Drainage Basin of River Broel, Germany. In: Kalma, J.D., Sivapalan M.: Scale Issues in Hydrological modelling, Wiley.

Glowinski, R. (1984): Numerical Methods for Nonlinear Variational Problems. Springer, New York.

Hornung, U. (1992): Applications of the Homogenization Method to Flow and Transport in Porous Media. In: Shutie, X. (Ed.): Summer School on Flow and Transport in Porous Media. Beijing, China, 8–26 August 1988. World Scientific, Singapore, 167–222.

Johanson, R.C., J. C. Imhoff, J.L. Kittle, A.S. Donigan, and C. Anderson–Nichols (1984): Hydrological Simulation Program – Fortran (HSPF): User Manual for Rel. 8.0. EPA, Athens, Georgia.

Leavesley, G.H., R.W. Lichty, B.M. Troutman, and L.G. Saindon (1983): Precipitation Runoff Modeling System: User Manual. Water Resources Inv. Rep. 83–4238, Denver, Colorado.

Luckhaus, S. (1993): Mathematical Problems in Hydrology (in German). In: SFB 350: Wechselwirkungen kontinentaler Stoffsysteme und ihre Modellierung, Universität Bonn, 24–26.

Maidment, D.R. (Ed.) (1992): Handbook of Hydrology. McGraw-Hill, New York.

Pikul, M.F., R.L. Street, and I. Remson (1974): A Numerical Model Based on Coupled One–Dimensional Richards and Boussinesq Equations. Water. Resour. Res. 10:295–302.

Pikul, M.F., R.L. Street, and I. Remson (1975): Reply. Water. Resour. Res. 11:510.

Vachaud, G., and M. Vauclin (1975): Comments on 'A Numerical Model Based on Coupled One–Dimensional Richards and Boussinesq Equations' by M.F. Pikul, R.L. Street, and I. Remson. Water. Resour. Res. 11:506–509.

Erosional Development of Small Scale Drainage Networks

K. Helming[1], M. J. M. Römkens[2], S. N. Prasad[3], and H. Sommer[4]

[1] Center for Agricultural Landscape and Land Use Research (ZALF),
Müncheberg, Germany
[2] USDA–ARS National Sedimentation Lab., Oxford, MS, USA
[3] Dept. of Civil Engineering, University of Mississippi, USA
[4] Dept. of Landscape Ecology, Technical University München, Germany

Abstract. Drainage networks are usually determined for large scale river systems. Small scale drainage networks for upland eroding areas have rarely been studied. In this scale, drainage networks of surface runoff substantially affect soil erosion. The objective of this study was i) to explore the similarities between drainage networks of runoff on eroding surfaces and those of river systems and ii) to determine the interrelationships between drainage network development and soil erosion. In flume experiments sequences of simulated rainstorms and overland flow were subjected to soils of initially different surface configurations. Before and after each rainstorm and overland flow test, digital elevation maps (DEMs) of the soil surface were generated using a laser scanner with 3 mm grid spacing. Drainage networks were determined from the DEMs and characterized with Horton's ratios, fractal characteristics, and with single stream properties like gradient, sinuosity, and orientation.
Horton's ratios indicated convergence and organization for all determined networks. When expressed with Horton's ratios and fractal characteristics, drainage networks of runoff on eroding surfaces were similar to those of river systems. Initially different network configurations yielded different erosion values but resulted in similar network characteristics at the end of the rainstorm and overland flow experiments. Raindrop detachment, clod destruction, and microrelief changes were identified as important mechanisms of network configuration and stream property changes during the rainstorm and erosion events. The network changes led to network structures that resulted in continuously decreasing soil erosion values. The results support the idea of optimization in drainage network development.

Introduction

Drainage networks, usually determined in the scale of river systems, often exhibit a striking degree of similarity. This phenomenon has been of interest to geomorphologists, who have tried to describe drainage networks and understand principles of network development. In the seminal work of Horton (1945), bifurcation and stream length ratios were introduced that describe the scaling properties of drainage networks. These properties provide useful links between drainage patterns and basin hydrology. Developed as empirical parameters, Horton's achievements were found to apply to a wide range of spatial scales and catchment configurations (Rodriguez–Iturbe and Valdez 1979, Rosso 1984). Even the more recent approaches of describing drainage networks with fractal characteristics can be related to Horton's ratios (Tarboton et al. 1988, Beer and Borgas 1993, La Barbera and Rosso 1989).

The often reported regularity of drainage networks, the universality of Horton's ratios and fractal characteristics raise questions about the principles of network evolution. For example, Rinaldo et al. (1992) and Rodriguez–Iturbe et al. (1994) suggested the theory of "self organization" as being the fundamental principle of network development. According to this theory, the evolution process of networks toward optimal configurations is triggered by the minimization of the total rate of energy dissipation in the system. The optimal network configuration is characterized by high fractal dimensions. Therefore, networks would evolve automatically to configurations exhibiting a fractal dimension of about 2, irrespective of the initial catchment configuration. Rinaldo et al. (1992) and Rodriguez–Iturbe et al. (1994) confirmed this theory on the basis of mathematical simulations of network evolution and they concluded that the theory provides a clear link between the patterns and the dynamics underlying the growths of networks. According to this idea, the fractal dimension serves as a measure of the degree of network organization and helps to forecast the further development of the network.

The similarity of drainage networks at different scales suggest that small scale runoff patterns on eroding upland areas might also have similar characteristics. Small scale runoff patterns have rarely been studied. Instead, the majority of studies of upland erosion have focused on the processes of runoff and erosion, which often are separated into interrill and rill components. Commonly accepted concepts like hydraulic shear (Foster et al. 1977) or stream power (Rose et al. 1983) are available to describe interrill and rill processes. Most applications of these concepts have been carried out in an one–dimensional context, in which the effect of spatially varying flow patterns on flow hydraulics were not considered (Nearing et al. 1989). The few attempts to describe the patterns of interrill runoff as well as the convergence of rills include the works of Scoging (1992), Zhang and Cundy (1989), and Wilson (1993). Wilson (1993) expanded on the idea to describe runoff patterns in terms of drainage networks. He proposed that interrill flow could be characterized with small flow paths that converge with other flow paths to form flow concentrations that converge into rills. According to this concept, a rill is defined as a flow path of a certain order. The approach of describing runoff patterns in terms of drainage networks has the advantage, that the tools needed for network characterization like Horton's ratios and fractal characteristics are already available. In contrast to the large scale process of river system evolution, the evolution of runoff patterns on small upland areas takes place within a reasonable short time span during a rainstorm or a sequence of rainstorms. The interrelationship between runoff patterns and erosion processes can therefore be studied in experiments. If drainage networks of runoff exhibit similar spatial configurations as drainage networks of river systems, it could be inferred that similar dynamics underlie network evolution on small scales. Then, the theories and principles of network evolution could be tested in experimental simulations.

The objective of our studies was to determine the development of small scale drainage networks for different initial conditions of eroding surfaces using flume studies with simulated rainstorms. Specifically, we wanted to test the hypothesis of a similarity between drainage networks of runoff on upland eroding surfaces and those of river systems, and to determine what the interrelationships are between drainage network development and soil erosion.

Materials and Methods

Equipment

Studies consisted of applying simulated rainstorm and overland flow to soils packed into a 3.7 m × 0.61 m × 0.23 m (length, width, depth) tilted flume. A schematic representation of the experimental setup is shown in Fig. 1. A multiple–intensity rainfall simulator consisting of three oscillating VeeJet nozzles (type 80150) was mounted above the flume. The simulator is similar in concept and design to the single nozzle simulator described by Meyer and Harmon (1979). For overland flow studies, an inlet tank was attached to the upper end of the flume. Water was admitted as uniform flow to the soil bed over a level, baffled edge on the downstream side of the tank. Soil beds prepared had different surface configurations. The surface elevations were measured with a removable laser microreliefmeter (Römkens et al. 1988) mounted on top of the flume. Its automatic movement into the longitudinal and transverse directions was driven by stepping motors and was software controlled. The soil surface area within the flume was digitized with a resolution of 3 mm in the horizontal directions and 0.25 mm in the vertical direction, respectively. A more detailed description of the experimental set up is given in Römkens et al. (1996).

Soil and Soil Bed Preparation

The Ap material of a Grenada loess soil (Glossic Fragiudalf) with a particle size distribution of 18 % clay, 79 % silt, 2 % sand, and with 1.1 % organic carbon was used in this study. The soil was obtained from a field at the North Mississippi Agricultural and Forestry Experiment Station (MAFES) near Holly Springs, MS., which had been in corn during the last two years. The soil bed was prepared in three stages. In the first stage, perforated drainage pipes were embedded into a 30 mm thick sand layer on the bottom of the flume to provide free drainage of percolating water. In the second stage, a 130 mm thick layer of air–dry soil, sieved to pass a 4 mm screen, was packed on top of the sand layer. Packing was conducted in a careful and systematic manner to achieve

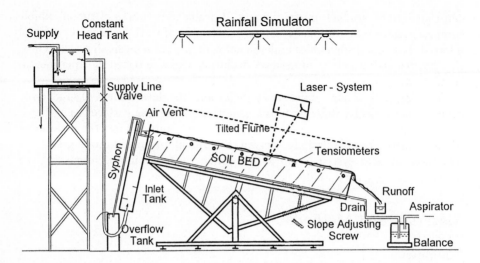

Fig. 1. Schematic diagram of the slope–adjustable flume with rainfall simulator and laser microreliefmeter.

a uniform density. In the third stage, the upper 70 mm of the soil bed was packed with air–dry soil, sieved through different screen sizes to obtain the desired surface configuration. Sieves of sizes 2 mm, 27 mm, and 56 mm were used to prepare smooth, medium rough, and rough surface conditions, respectively. With this procedure the surface configuration was determined by the diameter of the biggest clods of the surface material. The rough and medium rough surface conditions represented seedbed conditions commonly observed in the field. The smooth surface represented an artificial condition usually assumed in computational models describing overland flow.

Experimental Procedure

A total of four experiments were performed. Each experiment consisted of applying rain to the soil bed in a sequence of either seven rainstorms or of four rainstorms followed by two overland flow regimes on the initially air–dry soil. Slope steepness was 8 %, and the total amount of water applied was approximately 500 mm for each experiment. The first experiment was designed to study drainage network development for conditions of constant rainstorm intensity. The other three experiments were designed to study the effect of the initial surface configuration on network development and soil erosion. These three experiments consisted of applying a sequence of four rainstorms of decreasing intensity followed by a high overland flow rate. The specific design for each experiment is summarized in Table 1.

During the rainstorms and overland flow tests, samples were intermittently taken at intervals of 4 to 5 minutes at the flume outlet for gravimetric determinations of the runoff rate and sediment concentration. Before and after each

Table 1. Experimental design relative to surface configuration, amount and intensity of rainfall and overland flow. In all experiments, the total amount of water applied was 500 mm.

Exp.	Surface config.	Rainstorm and surface flow regime						
		test 1	test 2	test 3	test 4	test 5	test 6	test 7
1	rough	rain 50 mm 66 mm/h	rain 50 mm 66 mm/h	rain 50 mm 66 mm/h	rain 50 mm 66 mm/h	rain 50 mm 66 mm/h	rain 50 mm 66 mm/h	rain 200 mm 66 mm/h
2	rough	rain 45 mm 60 mm/h	rain 45 mm 45 mm/h	rain 45 mm 30 mm/h	rain 45 mm 15 mm/h	flow 86 mm 380 l/h	flow 233 mm 1034 l/h	
3	medium	rain 45 mm 60 mm/h	rain 45 mm 45 mm/h	rain 45 mm 30 mm/h	rain 45 mm 15 mm/h	flow 86 mm 380 l/h	flow 233 mm 1034 l/h	
4	smooth	rain 45 mm 60 mm/h	rain 45 mm 45 mm/h	rain 45 mm 30 mm/h	rain 45 mm 15 mm/h	flow 86 mm 380 l/h	flow 233 mm 1034 l/h	

rain = simulated rainstorm; flow = applied overland flow

test, digital elevation maps (DEMs) were made from laser microreliefmeter measurements of a 3.2 m × 0.6 m area within the flume. Grid distance was 3 mm resulting in DEMs with 219 096 data points each. The DEMs were used to determine drainage networks and its changes due to the rainstorm and surface flow regimes.

Drainage Network Determination

The procedures of obtaining drainage networks out of DEMs are based on the work of O'Callaghan and Mark (1984) and consist basically of three steps:

1. the determination of the flow direction for each cell within the DEM,
2. the determination of the number of cells that drain through each cell (contributing area), and
3. the assignment of an ordering system to the obtained streams.

In step 1 we used the one–directional drainage model by Jenson and Domingue (1988) in which the flow direction is assigned to that one of the eight neighbouring cells with the steepest descent. The pits (local depressions) were filled up by increasing the elevation until they drain. In step 2, the number of cells was counted that drain through each cell. From the resulting data set, streams could be defined as those cells that have a contributing area greater than a certain threshold. We used a threshold area of 25 cells (225 mm^2). In step 3, Strahler's ordering system was applied in which the exterior streams are the first order streams and higher order streams are defined as streams where two streams of the same order join (Strahler 1952).

Drainage Network Characterization

Three sets of parameters were used to characterize the drainage networks:

1. Horton's ratios (Horton 1945) that describe the network configuration,
2. stream characteristics that describe the individual components of the network, and
3. fractal dimensions that describe the network organization.

Horton's Ratios. Bifurcation ratio (Rb), length ratio (Rl), and drainage density (Dd) were determined for each drainage network:

$$Rb = \frac{N_{(i-1)}}{N_{(i)}} \tag{1}$$

$$Rl = \frac{L_{(i)}}{L_{(i-1)}} \tag{2}$$

$$Dd = \frac{\sum_{j=1}^{n} L_{(j)}}{A} \tag{3}$$

where $N_{(i)}$ is the number of streams of order i, $L_{(i)}$ is the median length of streams of order i, and A is the total drainage area. $\sum_{j=1}^{n} L_{(j)}$ is the sum of the flow lengths through all grids ($j = 1 \cdots n$) intersected by a stream. The flow length through grids with a straight lateral or longitudinal flow direction was equal to the grid length, and flow length through grids with diagonal flow direction was equal to the hypotenuse of the grid length. The average values of Rb and Rl were determined from the slopes of the straight lines resulting from plots of $\log N_{(i)}$ and $L_{(i)}$ versus order i respectively, according to Tarboton et al. (1988).

Stream Characteristics. For each single stream, the gradient, sinuosity, and orientation was determined as follows:

$$\text{Gradient}(\%) = \frac{100(h(G_{xy}) - h(G_{mn}))}{L} \tag{4}$$

$$\text{Sinuosity}(\text{m m}^{-1}) = \frac{L}{\sqrt{((m-x)w)^2 + ((n-y)w)^2}} \tag{5}$$

$$\text{Orientation}(°) = \tan^{-1} \left| \frac{x-m}{y-n} \right| \tag{6}$$

with

$h(G)$ = elevation of the grid G;

x, y = horizontal and vertical coordinates of the first grid in the stream;

m, n = horizontal and vertical coordinates of the last grid in the stream; and

w = grid length.

Fractal Characteristics. The fractal characteristics of the drainage network were determined from the network similarity dimension and the Richardson method of stream length according to Tarboton et al. (1988):

$$\text{Network similarity dimension } D_S = \frac{\log Rb}{\log Rl} \qquad (7)$$

The Richardson method consists of determining the total stream length L for various ruler sizes. The fractal dimension D_R is then equal to 1 minus the slope of the log–log plot of total stream length versus ruler size:

$$D_R = 1 - \frac{\log L_{(r)}}{\log r} \qquad (8)$$

where $L_{(r)} = N_{(r)} \cdot r$ is the total stream length measured with ruler size r, and N is the number of steps of size r. We calculated L with a ruler size ranging from 3 to 920 mm.

Results

Drainage Network Development (Experiment 1)

Changes in the surface configuration during the seven rainstorms were caused by soil detachment from and destruction of surface clods and by the development of one main flow path. Most of the smaller clods disappeared during the first 100 mm of rain, whereas the bigger clods were only gradually reduced in size. The position and shape of the main flow path was determined after 100 mm of rain. This flow path widened and grew upslope during the subsequent rainstorms (Fig. 2). The visualization of the upper one meter of the drainage networks showed random flow directions for the initial surface condition before rain was applied. During the rainstorm events, the streams seemed to become organized and directed toward the main stream (Fig. 3).

Horton's ratios indicated convergence and organization for all network configurations. All networks showed a maximum order of 5 or 6 and a log–linear relationship of numbers and lengths of streams versus stream order (Table 2). The development of the network during the application of rainstorms resulted in continuously increasing values of Rb and of the drainage density. This increase was the result of an increasing number of first and second order streams which increase was most appreciable during the first 100 mm of rain. The value of Rl was fairly constant during the first 200 mm of rain, but increased subsequently. The increase of Rl was due to the increased length of 6th order streams as compared to the 5th order streams. The average length of low order streams did not change during the entire experiment.

For each single stream, the gradient, sinuosity and orientation was calculated. The results yielded a high coefficient of variation, but trends could be discerned

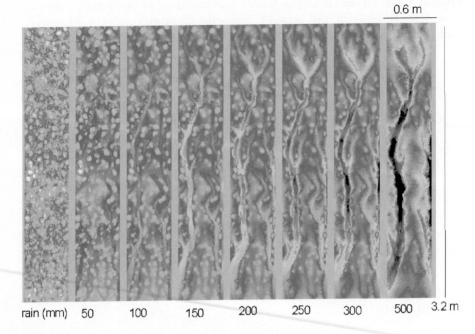

Fig. 2. Digital elevation maps of the soil surface profile of experiment 1: the initial condition (left) and after 7 successive rainstorms. Each map area was 0.6 m × 3.2 m and containing 220 000 elevation measurements.

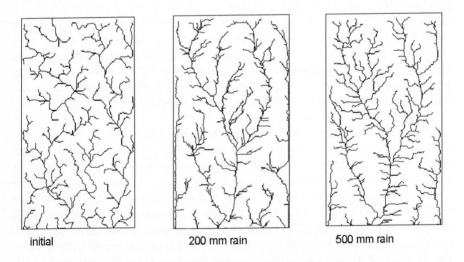

Fig. 3. Drainage networks determined from DEMs of the soil surface of experiment 1. The examples show the second and higher order streams of the upper 1m of the 0.6 m × 3.2 m area for the initial surface condition, after 200 mm of rain, and after 500 mm of rain.

Table 2. Horton's ratios, stream numbers, and stream lengths of drainage networks before and after each of a series of rainstorm tests of experiment 1.

cumulative		stream order						Horton's ratios		
rain (mm)		1	2	3	4	5	6	Rb	Rl	Dd
0	no	602	245	78	20	5	2			
	L	28	47	89	231	663	1317	3.28	2.23	28.9
50	no	866	279	78	18	4	1			
	L	27	41	95	302	1195	1141	3.95	2.36	35.2
100	no	1032	309	76	21	4	2			
	L	28	39	86	190	268	2254	3.67	2.26	38.6
150	no	1081	344	74	18	5	1			
	L	30	39	62	460	1043	1188	4.06	2.37	40.2
200	no	1115	338	75	18	3	1			
	L	30	42	85	431	1992	508	4.26	2.19	41.5
250	no	1088	369	73	15	3	1			
	L	30	42	65	410	500	3154	4.29	2.53	41.8
300	no	1132	358	80	15	2				
	L	31	44	57	558	3087		4.86	3.24	43.0
500	no	1013	388	81	16	3	1			
	L	31	51	57	247	546	3115	4.28	2.47	43.6

no = number of streams, L = median stream length (mm)

that distinguished these characteristics for different order streams (Fig. 4). Given the often detected skewness towards higher values in the density distributions of the stream characteristics, we calculated median values to determine the average stream characteristics as a function of stream order.

The average sinuosity of streams in the network obtained before the application of rain from the initial surface condition ranged between 1.3 and 1.4 with the higher values for the high order streams. The rainstorm regime led to a more direct routing of the streams toward the nearest junction resulting in sinuosity values of 1.2 to 1.3 (Fig. 4a). The average stream gradient was initially 14.2% for the first order streams and decreased to 9.3% during the first 100 mm of rain. The average gradient of the 4th and higher order streams was lower than the flume gradient of 8% (Fig. 4b). The rainstorm series led to decreased differences between the gradients and sinuosities of streams of different order. Most changes in the stream gradients and sinuosities happened during the first 200 mm of rain.

Stream orientation averaged initially about 40° deviation from the vertical (flume downslope direction) for first to third order streams. The 4th to 6th order stream orientation averaged initially between 10° and 25° from the vertical. The rainstorms mainly affected the orientation of the first to third order streams resulting in average values of 50° in the network after 500 mm of rain (Fig. 4c). Regarding the stream orientations, the different order streams could be grouped

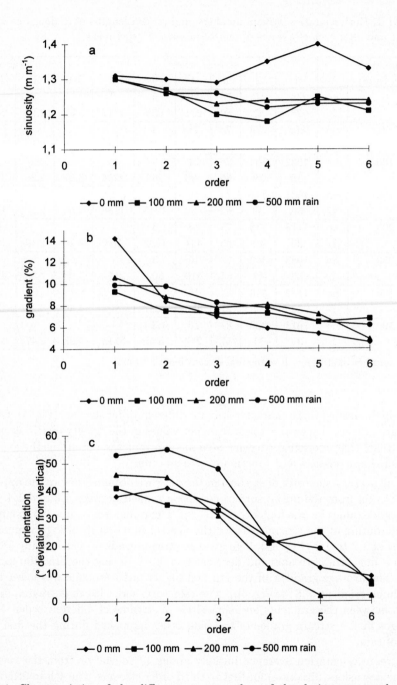

Fig. 4. Characteristics of the different stream orders of the drainage network after successive rainstorm events in experiment 1 (median values). a: stream sinuosity, b: stream gradient, c: stream orientation.

into two classes: first to third order streams with orientation approximately diagonal to the flume slope direction, and 4th to 6th order streams with orientations more or less parallel to the flume slope direction. The higher order streams (4th to 6th) seemed therefore to represent the local main streams, whereas the low order streams (first to third) represented the tributaries. In contrast to the changes in stream number, stream sinuosity, and stream gradient, most of the orientation changes took place during the application of the last 300 mm of rain.

Fractal characteristics of the drainage networks were determined to describe the degree of network organization. The log–log plots obtained from the Richardson method of drainage length (Fig. 5) show a smooth slope region for small, and a steep slope region for large ruler sizes, respectively. According to Tarboton et al. (1988) the smooth slope region corresponds to the sinuosity of individual streams, and the steep slope region corresponds to the branching characteristics of the network. The fractal dimensions D_R were derived from the slope of the linear regression of the log–log transformed data using Eq. (8). The individual streams yielded fractal dimensions D_R between 1.02 and 1.1 for all measured networks. The networks exhibited fractal dimensions D_R ranging from 1.75 for the initial surface configuration to 2.04 after 500 mm of rain. Thus, single stream characteristics and network branching characteristics could be clearly distinguished indicating a high degree of organization of the drainage networks. All networks, even that describing the initial condition with no rain, had similar fractal characteristics. Changes due to the rainstorm events were low.

The method of determining the fractal dimension D_S consisted of calculating the network similarity dimension on the basis of Horton's ratios (Eq. 7). This method allowed the differentiation of the networks obtained after successive rainstorms (Fig. 6). D_S increased to a value of 1.85 during the early rainstorms, but subsequently decreased and finally increased again. The highest value for D_S was obtained after 200 mm of rain which coincided with the completion of stream property changes, whereas the later changes of D_S coincided with the most intense changes of low order stream orientations.

The results from soil loss measurements are presented in Fig. 7. The sediment concentration increased during the first 100 mm of rain and subsequently decreased gradually to about one third of the maximum values. The first period of high sediment concentrations could be explained by soil detachment and clod destruction due to raindrop impact. The second period of decreasing sediment concentration was probably determined by an increase in the surface seal stability on one hand, and an increase in the stability and effectiveness of the drainage network on the other hand.

Effect of Surface Configuration on Network Properties and Network Changes (Experiments 2–4)

Three different surface conditions were tested in experiments 2 to 4. The rough and medium rough surface configurations simulated field surfaces after seedbed preparation. They only differed in the maximum clod size. The smooth configuration simulated an artificial uniform surface.

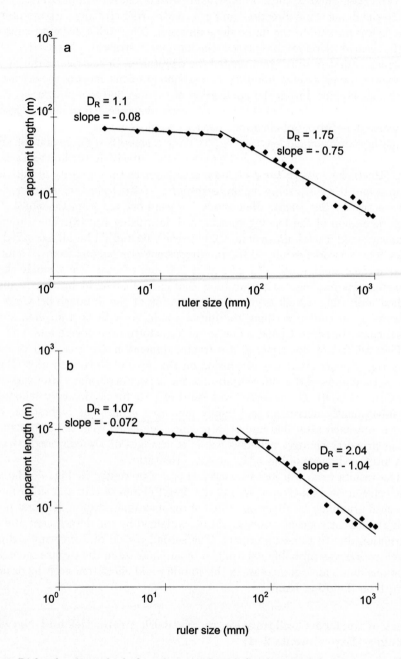

Fig. 5. Richardson's method of total stream length for the initial drainage network (a) and the drainage network after 500 mm rain (b) in experiment 1.

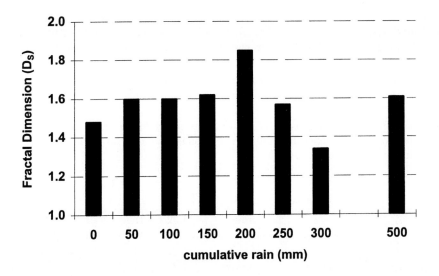

Fig. 6. Fractal dimension D_S derived from networks of the soil surface before and after each rainstorm event for experiment 1.

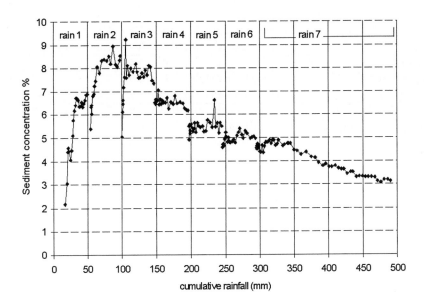

Fig. 7. Sediment concentration as a function of cumulative rainfall during a successive series of rainstorms in experiment 1.

The visualization of the upper one meter of the drainage networks showed a similar, random directed pattern of streams for the initial situation of the rough and medium rough surfaces (Fig. 8). Gradients, sinuosities and orientations of the streams were similar for these surfaces and were comparable to those of the network in experiment 1 (Table 3). The drainage density was 12 % higher for the medium rough than for the rough surface configuration (Table 4).

Table 3. Stream characteristics of the drainage networks for three different surface configurations before rain, after 180 mm of rainfall, and after an additional (702 l) of overland flow (Experiment 2–4).

	test situation	surface configuration	stream order					
			1	2	3	4	5	6
stream sinuosity	initial situation	rough	1.32	1.31	1.31	1.42	1.63	
		medium rough	1.34	1.34	1.34	1.43	1.62	1.61
		smooth	1.21	1.16	1.13	1.13	1.10	1.08
	after 180 mm rain	rough	1.27	1.23	1.21	1.18	1.19	1.15
		medium rough	1.27	1.22	1.18	1.17	1.14	
		smooth	1.24	1.19	1.15	1.13	1.16	1.20
	after 702 l overland flow	rough	1.33	1.28	1.30	1.22	1.20	
		medium rough	1.32	1.29	1.27	1.25	1.25	
		smooth	1.30	1.27	1.26	1.26	1.26	
stream gradient	initial situation	rough	11.9	8.8	5.8	5.0	5.6	
		medium rough	9.8	7.9	5.8	5.6	4.7	5.0
		smooth	6.6	6.9	7.2	7.3	7.2	7.4
	after 180 mm rain	rough	8.0	7.5	7.5	7.9	6.8	6.8
		medium rough	8.0	8.0	7.9	8.0	6.8	
		smooth	7.8	7.8	7.4	8.6	8.5	6.8
	after 702 l overland flow	rough	10.3	8.8	7.7	8.5	6.8	
		medium rough	10.4	9.4	10.3	9.1	5.8	
		smooth	9.8	9.0	11.2	9.5	3.6	
stream orientation	initial situation	rough	38	37	32	19	3	
		medium rough	38	32	35	22	7	10
		smooth	9	9	9	9	8	1
	after 180 mm rain	rough	26	25	20	24	5	4
		medium rough	23	21	19	22	3	
		smooth	12	13	12	10	7	5
	after 702 l overland flow	rough	41	35	33	22	5	
		medium rough	38	33	27	25	6	
		smooth	30	28	26	12	1	

The smooth surface had initially different network characteristics. In this case, the stream patterns could visually be described by approximate parallel

rough surface configuration

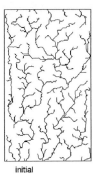
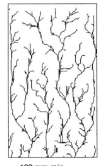
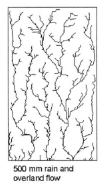

initial   180 mm rain   500 mm rain and
                                      overland flow

medium rough surface configuration

smooth surface configuration

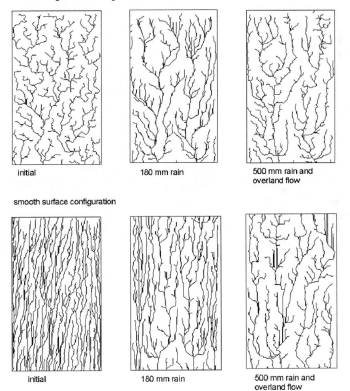

initial   180 mm rain   500 mm rain and
                                      overland flow

initial   180 mm rain   500 mm rain and
                                      overland flow

Fig. 8. Drainage networks obtained from DEMs of the soil surface profile of experiments 2, 3, and 4. The examples show the second and higher order streams of the upper 1 m of the 0.6 m × 3.2 m area for the initial surface condition, after 180 mm of rain, and after 500 mm of rain and overland flow.

Table 4. Drainage network characteristics and soil losses for three different surface configurations (Experiment 2–4).

	initial situation			after 180 mm rain			after 702 l overland flow		
	rough	medium rough	smooth	rough	medium rough	smooth	rough	medium rough	smooth
Dd	30.7	34.5	67.1	43.4	47.9	55.9	38.1	42.2	46.6
Ds	1.19	1.49	1.91	1.60	1.49	1.86	1.30	1.43	1.55
Soil loss (kg m^{-2})				3.3	2.9	1.4	2.9	4.3	9.6

Dd = Drainage density (m m^{-2}), Ds = Fractal dimension

streams with a high density (Fig. 8). The stream sinuosities ranged between 1.1 and 1.2. The average orientation of all streams was around 9°. The different order streams had gradients that were all smaller than the flume gradient of 8 % (Table 3). The drainage density was 67 m m^{-2} which was about twice as high as the drainage density for the soil bed with the rough surface configuration. Despite the near–parallel orientation of the streams, the network for the soil bed with the smooth surface configuration reached a stream order of 6 (Table 3).

After 180 mm of rain, the rough and medium rough surface configurations had similar soil losses of about 3 kg m^{-2}, while the smooth surface configuration yielded a soil loss of 1.4 kg m^{-2}. This difference in soil loss was attributed to differences in flow properties between these cases. The smooth surface had uniform flow, while spatially varied flow existed on the rough and medium rough surfaces. Evidently, the spatially varying flow distribution on the rough and medium rough surface configurations had places with high flow depth and flow velocity, where most soil erosion took place.

The soil loss rates obtained from the tests with 702 l overland flow were 9.6 kg m^{-2} for the initially smooth surface configuration. This soil erosion rate was two and three times that for the initially rough and medium rough surface configuration, respectively. The high soil loss rates during the overland flow regimes for the initially smooth surface were accompanied by appreciable changes in the flow distribution and stream properties for this surface.

The rainstorm and overland flow regimes resulted in decreased differences in the stream characteristics for the three surface configurations. A smoothing of the initially high surface roughness took place on the rough and medium rough surfaces resulting in decreased values of stream gradients and stream sinuosities, and an increase in the drainage densities. On the smooth surface, the rainstorm provided an increased variation in surface elevation of the initially uniform bed resulting in increased values of the stream gradients and stream sinuosities. The overland flow regime led to flow concentration on all three surface configurations as indicated by decreased drainage densities. The flow concentration was accompanied by increased variations of stream gradients and stream sinuosities and by the adjustment of the low order stream orientations towards a diagonal direction.

The drainage network development on the rough and medium rough surfaces resulted into an increase in the fractal dimensions D_S to 1.6 and 1.5, respectively during the rainstorm sequence, which then decreased to values of 1.3 and 1.43, respectively during the overland flow tests. On the smooth surface network development led to continuously decreasing values of the fractal dimension D_S from 1.91 initially to 1.55 after surface flow.

After the rainstorm and overland flow events, all networks had similar characteristics. The changes of the network and stream properties were most apparent for the initially smooth surface. This case also yielded the highest soil loss as compared to the other surface configurations. Thus, the initially different network configurations led to different erosion responses and different degrees of network development, which resulted in similar network patterns at the end of the experiments.

Discussion

The development of drainage networks on eroding surfaces was determined

1. to show that drainage networks of runoff on eroding surfaces may be comparable to those of river systems, and
2. to study the interrelationships between drainage network development and soil erosion.

All networks that were generated in this study could be described with Horton's ratios with maximum orders of 5 or 6. Horton (1945), investigating 11 networks draining areas of 10 km^2 up to 10^3 km^2, reported values for the maximum stream order of 4–7, Rb of 2.2–3.9, and Rl of 1.8–2.7, respectively. Rosso et al. (1991) studied drainage networks of up to 10^4 km^2 in size and found values for Rb ranging from 2.7–4.7 and for Rl from 2.0–2.5, respectively. The reported values are within the same range as those obtained in this study. Therefore, bifurcation and length scale properties of networks representing runoff on eroding surfaces seem to be comparable to those of networks representing river systems.

The reported values of fractal dimensions D_S for river systems ranged between 1.5 and 2.1 (Helmlinger et al. 1993, Marani et al. 1991, Nikora 1994). In applying the Richardson method, we obtained strikingly similar drainage length – ruler size relationships as those described by Tarboton et al. (1988) for several river systems. They measured fractal dimensions D_R of about 1.05 for single streams and about 2 for networks. Wilson and Storm (1993) found similar values for small scale drainage networks. The similarities in values for Horton's ratios and fractal characteristics between networks of different scales have repeatedly been postulated (Rinaldo et al. 1992, Nikora 1994). The results of our study suggest that these similarities extend to the scale of upland erosion areas. Thus, the flow patterns of interrill runoff exhibit convergence and regularity in network structure.

The drainage densities determined in this study were one order of magnitude larger than those of river systems (Horton 1945, Rosso et al. 1991). On one hand, the high drainage densities may reflect a low degree of flow concentration present on upland erosion areas (Montgomery and Dietrich 1992). On the other hand, the drainage density values are highly dependent on the resolution of the DEM and on the size of the threshold area chosen for the determination of drainage networks (Beer and Borgas 1993, Helmlinger et al. 1993). By considering only the 4th and higher order streams in this study, thus assuming higher threshold areas, one obtains drainage densities of less than 10 m m^{-2}. The drainage density seems to be a network property which is not scale invariant, but which reflects scale specific processes on one hand, and which depends on the measurement conditions like resolution and threshold definition on the other hand. The comparison of drainage densities determined for networks of different scales and measurement conditions must therefore be treated with caution.

In erosion studies, a threshold level is usually assumed to be a flow regime below which diffusive processes such as rain splash dominate and sheet flow exists, and above which processes of incision dominate and convergent flow properties prevail (Horton 1945, Kirkby 1980, Prosser and Dietrich 1995). The threshold depends on hydraulic tractive forces being able to overcome the erosion resistance of the surface material. This threshold has been expressed as the "critical shear stress" (Foster et al. 1977), or "threshold stream power" (Rose et al. 1983). The threshold concept thus implies a critical support area that is needed for incision to take place and for flow convergence to occur. Interrill and rill erosion processes are also distinguished by this threshold (Slattery and Bryan 1992). We measured convergent flow properties instead of sheet flow on interrill areas. Similar results were obtained by Wilson (1993). These results suggest that interrill and rill erosion might have similar flow patterns.

The method applied to flow network analysis consisted of directing the flow from one cell to only one of its eight neighbours. This method did not allow for the divergence of flow from one cell to several neighbours. Therefore, this approach overestimated flow convergence (Freeman 1991). Hence, the high degree of convergence in the networks obtained from the soil surfaces might to a certain extend be an artifact of the method used. On the other hand, a high resolution DEM was used in our study which helps to minimize this problem (Montgomery and Foufoula–Giorgiou 1993). Actually, this method was chosen because of its wide use. The methodology provided an opportunity to compare the network characteristics of runoff patterns with those of river systems that were determined with the same method. New algorithms of network determination have been developed that better describe divergent flow properties (Freeman 1991, Costa–Cabral and Burges 1994, Tarboton 1997) but are not yet widely implemented.

The networks obtained from the initial surface configurations before rain application were visually characterized by arbitrarily directed streams for the rough surfaces, and by almost parallel oriented streams for the smooth surface. The visual observations were supported by the measured stream characteristics that yielded a high degree of variation for the rough surface cases. The smooth surface was characterized by streams that were all in directions around 9° from the flume slope direction. From these observations one might infer random numbers of bifurcation and length ratios and low values of fractal dimensions. This was not the case, however. Log–linear relationships of Horton's ratios were obtained from the networks based on their initial surface conditions indicating network organization. The fractal dimensions D_R obtained with the Richardson method also indicated a well organized drainage network for the initial surface condition. The discrepancy between the anticipated and the measured results might to a certain degree be attributed to artifacts inherent in the method of network determination like filling up local depressions and routing the flow. Thus, Horton's ratios and the fractal dimensions might yield unrealistically high degrees of network organization and therefore can only be used to compare networks determined by the same method. Other methods of network determination need to be used to test whether the method of network determination affected the results.

The development of the drainage network during the rainstorm sequence may be separated into two successive phases. The first phase took place during the first 200 mm of rain and was characterized by decreasing values of stream sinuosities and stream gradients. At the same time, the number of first and second order streams increased which led to increased drainage densities. The second phase took place during the subsequent 300 mm of rainfall and was characterized by a shift of low order streams towards higher order streams.

During the first phase, soil erosion rates first increased and decreased gradually after 100 mm of rain. Increasing soil erosion rates during the early stages of a rainstorm over a dry, cloddy surface have often been observed and can be explained by the breakdown of surface clods upon raindrop impact (Bradford et al. 1987). After reaching maximum values, the erosion rate usually levels off to steady state values which are controlled by the detachment resistance of the surface seal (Moore and Singer 1990). Most of the first and second order streams developed during the first 100 mm of rain. Naturally, those areas on the surface that were initially covered with clods caused flow divergence during the early stages of the rainstorm tests. The breakdown of soil clods by raindrop impact resulted primarily into the development of low order streams. Further reduction in the surface microrelief due to raindrop impact led to decreased gradients and sinuosities of the low order streams. Clod destruction and reductions in the microrelief seemed therefore to be the main reasons for network changes during this first period of high erosion rates. The stream number and stream property changes resulted in increased fractal dimensions D_S yielding the highest value after 200 mm of rain. According to the theory of network self organization, a high fractal dimension would indicate a high degree of network organization (Rinaldo et al. 1992). Thus, high soil detachment and transport rates led to stream property changes which resulted in increased network organization.

In the second phase of network development during the later 300 mm of rain, the soil erosion rates decreased constantly. This result was somewhat unexpected, since continuous rain usually results into increased pore water pressures and a decreased resistance of the surface material to detachment, that would enhance the erosion rates (Römkens et al. 1996). Moreover, long lasting rainstorms often promote headcut incision and rill development which would increase soil loss due to increased stream power by rill flow (Rauws and Govers 1988). We measured constant infiltration rates of about 10 mm h^{-1} which must have continuously increased the subsurface soil water pore pressure. Additionally, the shift of first to third order streams toward 4th to 6th order streams indicated the increased importance of the high order main streams within the network system. The stream order adjustments were probably due to lowered elevations of the main stream beds, which in turn was indicative of rill flow. However, the stream power did not increase. In fact, the decreasing erosion rates suggest that the network seemed to develop a somewhat optimized runoff pattern that maintained the stream power below a critical threshold and drained the area with less soil detachment. Howard (1990) studied the development of optimal drainage networks in computer simulations by minimizing the total stream power within the network. He observed dynamics in stream generation and stream orientation adjustments that were similar to our measurements. Rinaldo et al. (1992) also studied the development of optimal drainage networks with computer simulations in which the local and global rate of energy dissipation was minimized. The main factors of energy dissipation are by friction (internal and between media) and by soil detachment. Thus, the optimization process would lead to flow concentrations that minimize friction, and the degree of flow concentration would be controlled by the threshold stream power to minimize soil detachment. Rinaldo et al. (1992) defined this process of minimizing the rate of energy dissipation in terms of "self organization". He postulated that networks yielding fractal dimensions near 2 are the result of network self organization. In our study, the fractal dimensions seemed to be less suitable for determining the degree of network organization. The values of D_R did not change appreciably during the entire experiment, and the values of D_S first increased but decreased after 200 mm of rain. However, the regularity of stream properties as well as the soil erosion dynamics of this study seemed to indicate the relevancy of the optimization process for runoff, thus supporting the idea of self organization.

The optimization phenomenon might also explain the fact that the development of the initially different networks of experiments 2, 3, and 4 resulted in similar network configurations at the end of the experiments. The rainstorm and overland flow events led to different erosion dynamics on these initially different surface configurations. The smooth configuration was characterized by uniform, almost parallel streams covering the entire surface. These small parallel streams yielded low soil loss rates. Once flow concentration was set up, the streams completely changed their configuration. These changes were accompanied by high soil loss rates. In contrast, the randomly directed streams on the rough surfaces reoriented in a downward direction towards regularity during the rainstorm events. The high overland flow rates resulted in minor changes of the stream properties and in lower soil loss rates when compared to the case of the smooth surface configuration. Thus, the different erosion responses on the different surface configurations were the result of stream property differences in network configuration. This suggests the notion that network configuration properties could be used more advantageously in describing runoff patterns and soil erosion. These experiments were a first attempt to study the processes involved in network development. Future work is needed based on experiments with different boundary conditions and different soil properties to test the concept of optimization principles involved in network development.

Summary and Conclusions

The development of drainage networks describing runoff patterns on eroding surfaces was determined in flume studies with simulated rainstorm and overland flow tests. The characterization of the drainage networks included the determination of Horton's ratios, fractal characteristics, and single stream properties.

The small scale drainage networks of runoff on eroding surfaces were found to have similar characteristics to those of river systems when expressed by Horton's ratios and fractal dimensions. The drainage networks were well organized and indicated a high degree of flow convergence. However, the method applied to determine the networks was found to imply the risk of overestimating flow convergence.

The rainstorm and erosion events led to the development of the networks with regular structures. Raindrop detachment, clod destruction and microrelief leveling were found to be the mechanism of changes of stream gradients and stream sinuosities. The decrease in the elevation of the main stream beds resulted in the adjustment of low order stream orientations towards the high order main streams. Stream property and stream configuration changes led to optimized network structures that resulted in continuously decreasing soil erosion values. Initially different surface configurations yielded different erosion rates and resulted in similar stream characteristics at the end of the rainstorm and overland flow experiments. The characterization of runoff patterns in terms of drainage networks were found to prove useful to describe the interrelationships between runoff patterns and erosion dynamics in small upland areas.

The following questions need further testing: to what extent does the method of determining drainage networks affect the network properties and the degree of flow convergence? Does the development of drainage networks represent an optimization phenomenon in which the total stream power and the rate of energy dissipation is minimized?

References

Beer, T., and M Borgas (1993): Horton's laws and the fractal nature of streams. Water Resources Research 29:1475–1487.

Bradford, J.M., J.E. Ferris, and P.A. Remley (1987): Interrill soil erosion processes: I. Effect of surface sealing on infiltration, runoff, and soil splash detachment. Soil Sci. Soc. Am. J. 51:1566–1575.

Costa–Cabral, M.C., and S.J. Burges (1994): Digital elevation model networks (DE-MON): A model of flow over hillslopes for computation of contributing and dispersal areas. Water Resources Research 30:1681–1692.

Foster, G.R., L.D. Meyer, and C.H. Onstad (1977): An erosion equation derived from basic erosion principles. Transaction of the ASAE 20:678–682.

Freeman, T.G. (1991): Calculating catchment area with divergent flow based on regular grid. Comput. Geosci. 17:413–422.

Helmlinger, K.R., P. Kumar, and E. Foufoula–Georgiou (1993): On the use of digital elevation model data for Hortonian and fractal analysis of channel networks. Water Resources Research 29:2599–2613.

Horton, R.E. (1945): Erosional development of streams and their drainage basins; hydrophysical approach to quantitative morphology. Bulletin of the Geological Society of America 56:275–370.

Howard, A.D. (1990): Theoretical model of optimal drainage networks. Water Resources Research 26:2107–2117.

Jenson, S.K., and J.O. Domingue (1988): Extracting topographic structure from digital elevation data for geographic information systems analysis. Photogrammetric Engineering and Remote Sensing 53:1593–1600.

Kirkby, M.J. (1980): The stream head as a significant geomorphic threshold. In: Coates, D.R. and J.D. Vitek (eds): Thresholds in geomorphology. Allen and Unwin, London.

La Barbera, P., and R. Rosso (1989): On the fractal dimension of stream networks. Water Resources Research 25:735–741.

Marani, A., R. Rigon, and A. Rinaldo (1991): A note on fractal networks. Water Resources Research 27:3041–3049.

Meyer, L.D., and W.C. Harmon (1979): A multiple intensity rainfall simulator for erosion research on row sideslopes. Transaction of the ASAE 22:100–103.

Montgomery, D.R., and W.E. Dietrich (1992): Source areas, drainage density and channel initiation. Water Resources Research 25:1907–1918.

Montgomery, D.R., and E. Foufoula–Georgiou (1993): Channel network source representation using digital elevation models. Water Resources Research. 29:3925–3934.

Moore, D.C., and M.J. Singer (1990): Crust formation effects on soil erosion processes. Soil Sci. Soc. Am. J. 54:1117–1123.

Nearing, M.A., G.R. Foster, L.J. Lane, and S.C. Finkner (1989): A process based soil erosion model for USDA–Water Erosion Prediction Project Technology. Transaction of the ASAE 32:1587–1593.

Nikora, V.I. (1994): On self–similarity and self–affinity of drainage basins. Water Resources Research 30:133–137.

O'Callaghan, J.F., and D.M. Mark (1984): The extraction of drainage networks from digital elevation data. Comput. Vision Graphics Image Pocess. 28:323–344.

Prosser, I.P., and W.E. Dietrich (1995): Field experiments on erosion by overland flow and their implication for a digital terrain model of channel initiation. Water Resources Research 31:2867–2876.

Rauws, G., and G. Govers (1988): Hydraulic and soil mechanical aspects of rill generation on agricultural soils. Journal of Soil Science 39:111–124.

Rinaldo, A., I. Rodriguez–Iturbe, R. Rigon, R.L. Bras, E. Ijjasz–Vasquez, and A. Marani, A. (1992): Minimum energy and fractal structure of drainage networks. Water Resources Research 28:2183–2195.

Rodriguez–Iturbe, I., and J.B. Valdes (1979): The geomorphic structure of hydrologic response. Water Resources Research 15:1409–1420.

Rodriguez–Iturbe, I., M. Marani, R. Rigon, and A. Rinaldo (1994): Self–organized river basin landscapes: fractal and multifractal characteristics. Water Resources Research 30:3531–3539.

Römkens, M.J.M., Y. Wang, and R.W. Darden (1988): A laser microreliefmeter. Transaction of the ASAE 31:408–413.

Römkens, M.J.M., S.N. Prasad, and K. Helming (1996): Sediment concentration in relation to surface and subsurface hydrologic soil conditions. Proc. of the 6th Federal Interagency Sedimentation Conference, Las Vegas, Nevada Vol. 2:IX 9 – IX 16.

Rose, G.W., J.R. Williams, G.C. Sander, and D.A. Barry (1983): A mathematical model of erosion and deposition processes. I. Theory for a plane land element. Soil Sci. Soc. Am. J. 47:991–995.

Rosso, R. (1984): Nash model relation to Horton order ratios. Water Resources Research 20:914–920.

Rosso, R.B., B. Bacchi, P. La Barbera (1991): Fractal relation of mainstream length to catchment area in river networks. Water Resources Research 27:381–387.

Scoging, H. (1992): Modelling overland–flow hydrology for dynamic hydraulics. In: Parsons, A.J. and A.D. Abrahams (eds): Overland Flow. UCL Press, 89–103.

Slattery, M.C., and R.B. Bryan (1992): Hydraulic conditions for rill incision under simulated rainfall: a laboratory experiment. Earth Surface Processes and Landforms 17:127–146.

Strahler, A.N. (1952): Hypsometric (area–altitude) analysis of erosional topography. Bulletin of the Geological Society of America 63:1117–1142.

Tarboton, D.G., R.L. Bras, and I. Rodriguez–Iturbe (1988): The fractal nature of river networks. Water Resources Research 24:1317–1322.

Tarboton, D.G. (1997): A new method for the determination of flow directions and upslope areas in grid digital elevation models. Water Resources Research 33:309–319.

Wilson, B.N. (1993): Small–scale link characteristics and application to erosion modeling. Transaction of the ASAE 36:1761–1770.

Wilson, B.N., and D.E. Storm (1993): Fractal analysis of surface drainage networks for small upland areas. Transaction of the ASAE 36:1319–1326.

Zhang, W., and T.W. Cundy (1989): Modeling of two–dimensional overland flow. Water Resources Research 25:2019–2035.

A Combined Conceptual Model for the Effects of Fissure–Induced Infiltration on Slope Stability

L. P. H. van Beek and Th. W. J. van Asch

The Netherlands Centre for Geo–ecological Research, Departement of Physical Geography, Utrecht University, The Netherlands

Abstract. In humid and subhumid Mediterranean environments the disruption of the vegetation cover by in particular shallow landslides limits the area in which erodible material is exposed to overland flow. On short temporal scales the hydrological system that allows for the generation of critical pore pressures or soil moisture conditions on the potential shear surface determines the occurrence of these landslides. Hence the relative contribution of landsliding to land degradation processes can be quantified in terms of its magnitude and frequency through a semi–physical hillslope model that links the relevant hydrological processes to a stability analysis. For the development of such a combined model for hydrology and slope stability a conceptualization of the process system is needed. Through the implementation of the resulting model in a GIS environment the effect of topography, to which the occurrence of landslides is intrinsically linked, can be incorporated. In addition, the spatial variation of hydrological and geomechanical parameters can be incorporated in the model. This is important since the occurrence and extent of shallow landslides is directly dependent on the net rainfall input in the hydrological system, as defined by the land cover of the area.

In this paper a conceptual model is presented that combines a description of the hydrology with an assessment of the slope stability. The conceptual framework of this model is based on field observations in the Alcoy area (SE Spain). In this area shallow landslides occur on steep, unsaturated slopes in marly deposits of Miocene age, at the boundary between regolith and bedrock (1–2 m depth). Given the low matric permeability of the marl it has been assumed that preferential flow along distinct sets of fissures by–passing slower matric percolation might account for the observed response time of landslides to rainfall events. The fissures in the regolith are either relict primary bedrock structures (discontinuities) or are formed by weathering, creep and shear; they are supplied with water by subsurface flow through the more permeable rootzone. With the combined slope stability model, which is programmed in the meta–language embedded in the PCRaster GIS package, a sensitivity analysis has been performed to assess the impact of fissure flow on the occurrence of landslides in a small catchment of 1.2 km^2 near Alcoy.

Introduction

Quantification of landslide occurrence in terms of magnitude and frequency is not only a valuable tool in hazard assessment, it also provides useful information on the relation of landsliding with other environmental processes (van Asch and van Steijn 1991, van Steijn 1996). This relationship is either direct if the redistribution of material affects the potential energy stored, or indirect when it affects the dynamics of the processes within the environmental system.

The temporal activity of landsliding is inherently determined by the frequency at which possible triggers occur. Landslides will be triggered when, due to a relative increase, the disturbing gravitational forces that act on a potential shear surface exceed the total of all forces resisting movement. Part of this resistance is formed by the intergranular friction that can be mobilized along the shear surface under the weight of the overlying material. The contribution of this frictional component to slope stability can thus be reduced if the weight or *effective stress* acting on the soil matrix in the shear surface is diminished. Hydrological triggers leading to slope instabilty affect the static equilibrium mainly by decreasing the effective stress, either through the generation of positive pore pressures or even by a decrease in matric suction on the potential shear surface (Anderson and Howes 1985, Fredlund 1987, Rahardjo et al. 1993).

Because of the short term fluctuations, ranging from event to seasonal scales, in the storage of the unsaturated and saturated zone on one hand, and because of the comparatively low recurrence intervals of critical rainfall events, the temporal activity of landslides is primarily controlled by the hydrological system (van Asch 1980, Anderson et al. 1988, van Asch and van Steijn 1991). The response time of a landslide to precipitation is for this reason dependent on the hydrological processes which influence the transmission of the precipitation input to the critical depth at a potentially unstable location. As a consequence different types of landslides have different hydrological systems and react at different temporal scales to the net precipitation input.

The prevalence of hydrological triggering mechanisms and the active erosion and deposition by overland flow give the occurrence of landslides a distinct place in the cascade of land degradation processes in Mediterranean environments (van Asch 1980). In humid and subhumid environments, where vegetation is not limited by water stress, shallow landslides in particular destroy the otherwise complete vegetation cover, after which material susceptible to erosion is exposed to overland flow. An assessment of the temporal activity of shallow landslides in such an environment can therefore be used to define the extent of the erosion prone area at different moments.

The temporal activity of shallow landslides is, in comparison to areas subject to more deep–seated landsliding, controlled by the travel time of the net precipitation input through the unsaturated zone of the regolith. For the transmission of the precipitation input, the presence of paths of preferential flow is of the utmost importance as these may allow water to by–pass the slower percolation through the unsaturated matrix (Beven and Germann 1982, Germann 1990, Bouma 1990). In the regolith, these preferential flow–paths comprise randomly distributed macropores formed by biological activity and sets of fissures which can either be relict primary structures (discontinuities) or tension cracks due to weathering, creep or shear. Sets of near–vertical fissures may penetrate the entire depth of the regolith and constitute a network with direct access to the potential shear surface. Accumulation of water in this fissure network will lead to slope instability in two distinct ways. On one hand, any water stored in the near–vertical fissures will increase the downslope component of the total weight of the soil above the shear surface, i.e. it will add to the disturbing forces leading to failure, while no additional effective stress is gained that might resist movement. On the other hand, the water level in the fissure network exerts a positive pressure at the bottom of the fissure; this will give rise to vertical drainage from the fissure into the underlying material, but – more effectively – it will also result in horizontal infiltration over the larger zone of contact into the saturated or unsaturated soil matrix between the fissures. In this manner, water can infiltrate directly at the critical depth where a loss of matric suction or the development of positive pore pressures might lead to instability. Since many landslides occur in less permeable deposits, this may be a likely mechanism to explain the observed rapid response time to the triggering rainfall event.

If, by means of a hydrological model, the afore–mentioned pressumed triggering mechanisms can be combined with a slope stability analysis, an assessment of the impact of fissure flow on the temporal activity of shallow landslides can be made. This hydrological model has to describe both unsaturated and saturated flow and the spatial redistribution of the precipitation input on a hillslope scale to model the matrix–fissure interactions succesfully. With regard to the processes in the unsaturated zone and in the fissure network, the temporal resolution of the net precipitation input must be adequate to draw meaningful conclusions on the influence of these processes on the temporal activity of landsliding. Under these considerations the modelling approach is by nature conceptual and deterministic; it requires a schematization of the relevant components of the entire process system which are described semi–physically.

In this paper a conceptual, combined hillslope model is presented which considers fissure flow and the consequent infiltration into the matrix. The assumptions underlying this model are based on field observations in the area near Alcoy (SE Spain) where shallow landslides occur on steep slopes in marly deposits of Miocene age. The model is programmed in the meta–language of the PCRaster GIS package and can used in combination with rasterized input maps and timeseries (PCRaster 1996, Wesseling et al. 1996). Implementation of such a combined hillslope model in a GIS–environment has several advantages. First of all, if based on a high resolution DEM, the effect of the topography can readily be incorporated on both local and higher scales. On the hillslope and sub–catchment scale, the implementation in a GIS offers the direct use of the available routing functions to define flow paths for saturated and unsaturated flow. On a local scale, the variations in slope angle to which the occurrence of landslides is intrinsically linked, are included and morphometric factors which are of importance for the occurrence of landslides can be analyzed (Carrara et al. 1995). Secondly, it is possible to include the spatial variation of the hydrological and geotechnical parameters. In a GIS environment this spatial variation can be linked dynamically to the patterns of land cover in the area which define the net precipitation input on which the occurrence and extent of shallow landslides is directly dependent.

Here the model has been used to perform a sensitivity analysis to assess the impact of fissure flow on the occurrence of shallow landslides in a small catchment of 1.2 km^2 near Alcoy.

Model Description

Conceptualisation of Matrix–Fissure Interactions

The combined model which consists of the hydrological and slope stability modules `fisswepp.mod` and `stabmod1.mod` respectively, is based on a conceptualisation of the unsaturated and saturated hydrological interactions in the fissured regolith of marly deposits in the area around Alcoy. In these marly deposits of Miocene age a pediment has formed during the Pleistocene. This pediment has been dissected by deep gullies or *barrancos*, probably since the Bronze Age (approx. 3000 BP; personal com. La Roca). On the steep side walls of these gullies shallow landslides are triggered at the contact between the regolith and bedrock (La Roca and Calvo–Cases 1988, La Roca 1991). Presently these shallow landslides with depths between 1 and 2 m, are studied in two small catchments – the *Barranco de Mollo* and the *Barranco de la Coloma* – in the municipality of Almudaina.

In the regolith of the study area, two distinct layers can be recognized. The first is a topsoil of maximum 0.5 m depth, which consists of a tillage layer or root zone under natural vegetation. This layer has a relatively high permeability because of the well–developed structure and texture under these types of land cover. The average saturated hydraulic conductivity of this layer, as measured by the inverse auger test, is 1.50 m d^{-1}. However, conductivities up to 4.6 m d^{-1} have been measured on well–rooted forest soils. This layer is formed on top of a less developed layer of weathered and remoulded marl with a soil of a clay–silt texture. The average saturated hydraulic conductivity for the matrix of this layer is 0.50 m d^{-1}. This regolith overlies the intact marl bedrock which with an average saturated hydraulic conductivity of 0.25 m d^{-1} is even less permeable. In the regolith near–vertical fissures can be recognized which extend over its entire depth. These fissures are partly relict primary structures, either discontinuities or bedding planes, or are partly cracks formed by weathering or under tension by creep or shear between shards of more intact material. Sets of these regulary spaced fissures form a network that is connected to the more permeable upper layer by the macropore network and by the roots that penetrate from the topsoil into the fissures. Given the low matric permeability of the marl bedrock and regolith these fissures are preferential flow paths which allow for triggering of the shallow landslides on shorter temporal scales than by matric percolation alone. In the conceptual model the two layers of the regolith and a buffer of bedrock are approximated by three layers with different hydrological and geomechanical characteristics.

From the third, bedrock layer water is lost to an infinite bedrock reservoir. In the hydrological model the input of precipitation into the first layer is saturation limited. In each layer both the percolation through the matrix and to the underlying layer are calculated and the difference is used to generate a perched water table. From the matrix, lateral, saturated outflow occurs if the water level in the fissure network is below the height of the perched water table in a layer. The accumulation of water in the fissure network may eventually lead to a gradient directed into the matrix. Under this gradient water from the fissure network can infiltrate into the matrix over both the saturated and unsaturated zone of contact between them. With the new hydrological conditions the slope stability is calculated at the end of every timestep, for which the lower boundaries of every layer are considered as potential shear surfaces. The spatial resolution used in the model is 10×10 m, the temporal resolution of the timeseries with precipitation is in days.

In order to describe the hydrological interactions between the matrix and the fissure network in a GIS–environment the distribution and properties of the different sets of fissures have to be quantified spatially. In most if not all occasions, however, such a complete inventarisation of the distribution and characteristics of the fissure network is not feasible. Moreover, an approach in which the fissure network is modelled explicitly is not consistent with most raster–based GIS packages for they use uniform pixel sizes. Introducing smaller sized pixels, in line with the dimensions of the fissures, is not a valid solution as it will slow calculations beyond practical applicability. For this reason the arrangement of the fissure network within the pixel is modelled implicitly, i.e. at a sub–pixel scale. For every layer the volume of the pixel is disaggregated in similar, average blocks of matrix, divided by fissures of equal size (Figure 1). With assumptions on the nature of the fissure network within the pixel, which should be based on field observations, the size of the average matrix block can be calculated with two variables: the volumetric fraction of the pixel volume taken by fissures or macropores, $X_{Fissure}$, and the mean effective aperture, $\hat{A}_{Fissure}(m)$, of the fissures in every layer. The effectiviness of the method is evidently limited by the assumptions made on the characteristics of the fissure network but it has some practicality; both the fraction occupied by fissures and the mean effective aperture can be dimensioned from field observations and can be used in a probabilistic analysis.

With the fraction $X_{Fissure}$ and the mean effective aperture $\hat{A}_{Fissure}$, all fissure–matrix interactions are calculated two–dimensionally at the sub–pixel scale using the mean radius, $R_{Mat}(i)$, between the fissure wall and the centerline of the average matrix block in every layer (Figure 2). This distance is calculated with the assumptions that

1. all fissures can be approximated by a space with a width equal to the mean effective aperture between two parallel surfaces;
2. all fissures are orthogonal and equally spaced;
3. all fissures extend over the total depth of the layer;
4. all fissures within the pixel are interconnected and not confined;

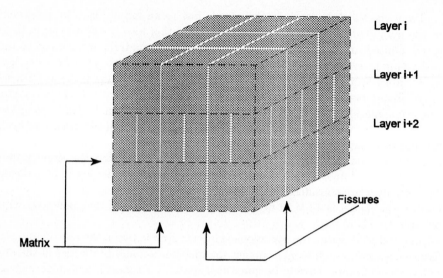

Fig. 1. Schematization of the fissure network.

5. the matrix in every pixel is surrounded by fissures.

It follows from the fourth assumption that the water level in a fissure exerts a pressure equal to the elevation head at its base. As a result of the last assumption all outflow from the matrix is stored first within the fissure network before it can be routed. The routing of water through the fissure network takes place over a Local Drainage Direction– or LDD–map which is derived from the topography and a friction factor defined by the fraction of fissures in the downstream pixels. If the difference in elevation between two or more downstream pixels is less than the regolith depth, the outflow is directed to the pixel with the highest fissure density. The actual amount of routing is dependent on a coefficient C – defined in the next section – which can be used as a calibration factor in modelling. When the coefficient C is set to zero, in which case no water will be transferred downslope through the fissure network, the saturated outflow from the matrix in excess of the fissure storage is transported to the downslope pixel over a LDD– map based on the topography only. An additional calibration factor to the model is a recharge parameter of the fissure network which describes the percentage of direct input due to rainfall and which can be modified to include precipitation excess or exfiltration from the matrix. This factor, however, is default set to zero which implies that the recharge of the fissure network is dependent on the saturated matrix outflow as described before.

Hydrology

As described in the conceptual framework the hydrological processes are calcu- lated separately in the hydrological model `fisswepp.mod`. An overview of the fluxes within and between the three layers of the model, the storage in the fissure network and the outflow to the infinite fourth reservoir are given in Figure 2. Because of the discretization of the model in both time and space a numerical solution is required to calculate the fluxes within and between the pixels. Given the relatively low temporal resolution of the model and the limitations of the PCRaster programming language, a simple approximation is used in which all gradients and hydraulic conductivities are defined as the average values. For the gradients these average values are equal to the half of the actual difference. For the hydraulic conductivity the average hydraulic conductivity as defined by Vauclin et al. (1979) is used, i.e. $k_{average}(i) = \sqrt{(k_{sat}(i) \, k_\theta(i))}$, in which $k_\theta(i)$ is the unsaturated conductivity belonging to the actual soil moisture content $\theta(i)$.

The input in the model is not limited by infiltration; if the net rainfall in a timestep exceeds the storage left in the first layer, a quantity equal to the free pore space, augmented with the percolation from the first to the second layer, is stored. The total matric percolation in each layer is calculated by the percolation rate $P_{Mat}(i)$ $(\mathrm{m\,d^{-1}})$, as used in WEPP (Lane et al. 1992)

$$P_{Mat}(i) = (\theta(i) - \theta_{FC}(i)) \left(1 - \exp(\frac{-\delta t}{t(i)})\right) \quad \text{if} \quad \theta(i) > \theta_{FC}(i) \tag{1}$$

and

$$P_{Mat}(i) = 0 \quad \text{if} \quad \theta(i) <= \theta_{FC}(i) \tag{2}$$

where $\theta(i)$ and $\theta_{FC}(i)$ are the respective soil moisture content for the current timestep and the soil moisture content at field capacity for layer (i) $(\mathrm{m^3\,m^{-3}})$, δt is the timestep and $t(i)$ is the travel time for soil moisture in excess of the field capacity through the layer, both given in days. The approach of the WEPP–model to calculate the percolation rate has been selected because of its simple parametrization and because it truncates percolation in any layer if the soil moisture content approaches field capacity. So, if the soil is at field capacity there is no percolation (Eq. 2), if the soil moisture content exceeds the field capacity, the travel time is calculated from the linear store equation, using the unsaturated hydraulic conductivity $k_\theta(i)$ $(\mathrm{m\,d^{-1}})$

$$t(i) = \frac{(\theta(i) - \theta_{FC}(i))(D(i) - H(i))}{k_\theta(i)} \tag{3}$$

where $D(i)$ and $H(i)$ are respectively the depth of layer (i) and the height of the perched water table (m). The unsaturated hydraulic conductivity $k_\theta(i)$ is calculated by the empirical relationship

$$k_\theta(i) = k_{sat}(i) \left(\frac{\theta(i)}{\theta_{max}(i)}\right)^{B(i)} \tag{4}$$

where $B(i)$ is equal to $-2.655/^{10}log(\theta_{FC}(i)/\theta_{max}(i))$ which forces $k_\theta(i)$ to approach $0.002\,k_{sat}(i)$ at field capacity. The actual percolation from layer (i) to

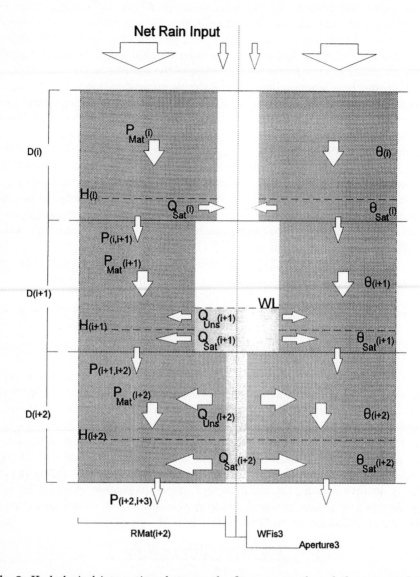

Fig. 2. Hydrological interactions between the fissure network and the saturated and unsaturated matrix in every layer; In the figure the fissure, of width $W_{Fis}(i)$, is represented by the blank area between two blocks of matrix (shaded) and filled to the level WL. The percolation rates through the matrix and to the next layer are given by $P_{Mat}(i)$ and $P(i, i+1)$ respectively, and result in a soil moisture content $\theta(i)$. The lateral fluxes over the saturated and unsaturated zones of each layer, as defined by the gradient over the distance $R_{Mat}(i)$ between the fissure wall and the center line of the matrix block, are denoted by $Q_{Sat}(i)$ and $Q_{Uns}(i)$. The perched water tables in each layer are represented by the dotted line, $H(i)$.

layer $(i + 1)$ is calculated as

$$P(i, i + 1) = P_{Mat}(i) \sqrt{1 - \frac{\theta(i + 1)}{\theta_{max}(i + 1)}} \qquad (5)$$

when no perched water table is present in layer (i), or

$$P(i, i + 1) = \frac{1}{2} H(i) \sqrt{k_{sat}(i + 1) k_{\theta}(i + 1)} \sqrt{1 - \frac{\theta(i + 1)}{\theta_{max}(i + 1)}} \qquad (6)$$

when a perched water table exists. The difference in actual and total matric percolation is used for the change in the height of the perched water table in layer (i). From the third layer the matric percolation is lost to an infinite reservoir with the same hydraulic properties but with an average hydraulic conductivity based on the actual $k_{\theta}(i)$ of the third layer and the unsaturated conductivity at field capacity.

The changes in the storage of the fissure network due to vertical drainage and routing are calculated independently from the matric flow in each timestep. Both are driven by the average gradient which is defined as half the piezometric head between the bottom and the water level in the fissure network of a pixel. The loss due to the vertical drainage is calculated similar to Equation 6. The routing through the fissure network, $Q_{Fissure}$ in $m^2 d^{-1}$, is calculated with the gravity store equation

$$Q_{Fissure} = C \, WatLevel \sqrt{\frac{1}{2} g \, WatLevel \tan(\alpha)} \qquad (7)$$

where C is the coefficient of performance, g is the acceleration due to gravity $(m\,s^{-2})$, $WatLevel$ is the water level in the fissure network (m) and α is the local slope angle $(°)$.

The matrix–fissure interactions are calculated by means of defining different gradients for the saturated and unsaturated zones of the matrix. The saturated gradient is the total difference in elevation head between the height $H(i)$ of the perched water table and the water level in the fissure network over the mean distance between the mid–point of an average matrix block, $R_{Mat}(i)$ - see Figures 1 and 2. The total flux over the saturated zone in $m^2 d^{-1}$ equals

$$Q_{Sat}(i) = \frac{1}{2} h_w k_{sat}(i) \frac{WatLevel - H(i)}{R_{Mat}(i)} \qquad (8)$$

where h_w is the minimal height of saturated contact between the matrix and fissure network. As can be seen in Equation 8, a positive flux denotes flow from the fissure into the matrix, whereas negative values denote drainage from the matrix into the fissures. When the water level is not in contact with the layer (i), the lower boundary of that layer is used to calculate the gradient for the saturated, lateral outflow from the matrix, with $h_w = H(i)$. For the calculation of $Q_{Sat}(i)$ both the water level in the fissure network and the height of the perched water table $H(i)$ from the previous timestep are used. For the gradient in the unsaturated zone a similar approach is used, except that in addition to the elevation head the matric suction in the current timestep, i.e. after the soil moisture content has been altered by matric percolation, is considered in the calculation of the gradient. The resulting flux $Q_{Uns}(i)$ is dependent on the average unsaturated conductivity and on the height of contact over the unsaturated zone which is $WatLevel - H(i)$. If the water level in the fissure network is not in contact with the unsaturated matrix, the unsaturated fissure wall is considered as a "no–flow boundary" and all unsaturated flow in the matrix is taken to be vertical.

Slope Stability

The output of perched water tables, soil moisture contents and the water level in the fissure network from the hydrological model is used by the stability model stabmod1.mod to estimate the safety factor SF as a measure for the slope stability. This safety factor is dimensionless and is derived from a static, limiting equilibrium analysis as the ratio between the resisiting and disturbing forces:

$$SF = \frac{F_{Resisting}}{F_{Disturbing}} \tag{9}$$

where:

$F_{Resisting}$ = the sum of the maximal frictional forces that can be mobilized to resist shear, and

$F_{Disturbing}$ = the sum of the downslope component of all the gravitational forces acting on the soil.

Hence, if $SF \geq 1$ the resisting forces are equal to the disturbing gravitational forces and the slope is thought to be stable; if $SF < 1$ the disturbing forces exceed the resisting frictional forces and the slope is likely to fail.

The assessment of the safety factor is made under the assumption of an infinite slope, i.e. there is no constraint on either side of the soil mass under consideration. The totals of the resisting and driving forces can for this reason be expressed as

$$F_{Resisting} = c' W_{Mat} + W' \cos(\alpha) \tan(\phi') \tag{10}$$

and

$$F_{Disturbing} = W \sin(\alpha) \tag{11}$$

respectively, where W is the total weight of the pixel (kN), W' is the effective weight of the matrix within the pixel (kN), W_{Mat} is the width of the matrix (m), α is the slope angle (°), c' is the effective cohesion (kPa) and ϕ' is the effective angle of internal friction (°). In the model the safety factor of each layer (i) is calculated with the base of that layer as the potential shear surface. In the analysis W is the total weight of the pixel above this layer and includes both the weight of the matrix, pore water and the water stored in the fissure network. The effective weight W' is the weight of the matrix above the potential shear surface minus the force exerted by the pore pressure acting on it. This pore pressure is – under steady conditions – equal to $H(i)\cos(\alpha)$. If no perched water table is present at the base of layer (i), the matric suction is used as the pore pressure in order to simulate the stabilizing effect it has on slope stability (Anderson and Howes 1985). This allows even slopes in non–cohesive soils steeper than the angle of internal friction ϕ' to be stable if the matric suction at the potential shear surface is high enough.

To obtain a better resolution of the distribution of the safety factor around the threshold of instability ($SF = 1$), all values have been forced within the total range between 0 and 2.

Data Requirements: Variables, In– and Output

For the parametrization of the combined hillslope model with the three conceptual layers a total of $9(i)+4$ state variables for geometry, hydrology and stability and $2(i)+1$ dynamic variables is required. For the input a timeseries with daily rainfall is needed, as well as a DEM, from which two state variables - the slope α and the pixel width - are derived. All data requirements are listed in Table 1.

The output of the combined model comprises the height of the perched water table $H(i)$, the soil moisture content $\theta(i)$ and the safety factor of each layer and the water level in the fissure network for every timestep. The total of rainfall excess and exfiltration to the surface in each timestep is reported as well.

In the stability model two maps are generated which are related to the temporal activity of landsliding; the first map reports the total number of days with instability, the second, which gives a better resolution for the less frequent instable pixels, reports the activity as the ^{10}log of that value, scaled over the total length of the period.

Model Results

With the combined slope stability model a sensitivity analysis has been performed to assess the impact of fissure flow on the occurrence of landslides in the small catchment – 1.2 km^2 – of the Barranco de la Coloma. The basis for this modelling exercise is formed by a timeseries of 60 days with net precipitation input – 108 mm divided over several events in the first 30 days of the period (see Figure 3) – and by a 10x10 m DEM of the catchment with 12303 pixels. All simulations are based on the same static parameters for each conceptual layer, which have been given depths of respectively 0.3, 0.7 and 1.0 m (Table 1). These layers are interpreted as the topsoil, the regolith of weathered marl and a buffer of bedrock. From the bedrock layer water is lost to an infinite bedrock reservoir.

The sensitivity analysis consists of 16 model runs that comprise 5 different fissure scenarios (see below) and 4 varying model settings:

Table 1. Data requirements and parametrization for the combined conceptual models `fisswepp.mod` and `stabmod1.mod`

		Description	Layer 1	Layer 2	Layer 3
General Input: - Net precipitation as a timeseries with a temporal resolution of days - High resolution Digital Elevation Model					
Static variables					
Geometry					
B	(m)	width pixel	10.0; derived from DEM		
α	(°)	slope angle	var.; derived from DEM		
$UL(i)$	(m)	upper limit	2.0	1.7	1.0
$X_{Fissure}$	$(\mathrm{m^3\,m^{-3}})$	fraction of fissures	dependent on scenario's		
$\hat{A}_{Fissure}$	(m)	mean effective aperture	0.002	0.002	0.002
Hydrology					
$K_{sat}(i)$	$(\mathrm{m\,d^{-1}})$	saturated hydraulic conductivity	1.50	0.50	0.25
$\theta_{sat}(i)$	$(\mathrm{m^3\,m^{-3}})$	porosity	0.50	0.48	0.40
$\theta_{FC}(i)$	$(\mathrm{m^3\,m^{-3}})$	soil moisture field capacity	0.29	0.23	0.23
Geomechanics					
$\gamma_{dry}(i)$	$(\mathrm{kN\,m^{-3}})$	dry bulk density	14.5	14.5	14.5
$c'(i)$	(kPa)	cohesion	20	0	0
$\phi'(i)$	(°)	angle of internal friction	38	38	45
Dynamic variables; initial settings					
$H(i)$	(m)	height of perched water table	0	0	0
$\theta(i)$	$(\mathrm{m^3\,m^{-3}})$	soil moisture content	dry initial conditions: $\theta_{FC}(i)$		
			wet initial conditions: 0.40	0.36	0.32
WL	(m)	water level in fissures	0	0	0
Calibration factors					
C	(-)	coefficient of performance of fissure network	0, 0.002^3		
R	(-)	direct recharge fissures by rainfall	0		
Total:					$11(i)$ +5 variables

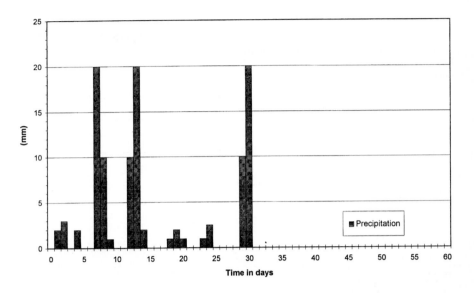

Fig. 3. Timeseries of 60 days with net precipitation input.

- 2 values for the coefficient of performance C of the fissure network, i. c. $C = 0$
  – there is no direct contact between the fissures of the different pixels – and
  C being the third power of the mean, effective aperture as an approximation
  of the "cubic law" (Snow 1968);
- two different initial soil moisture contents $\theta(i)$ for the successive layers, simulating the dry – or summer – conditions and the wetter conditions during
  winter. For the summer conditions, values are taken to be equal to the soil
  moisture content at field capacity ($\theta(i) = \theta_{FC}(i)$); for the winter conditions
  the average of $\theta_{FC}(i)$ and $\theta_{max}(i)$ is taken.

The fissure scenarios, all with a constant mean effective aperture of 0.002 m,
are:

1. no fissures in any layer;
2. both layer 1 and 2 have a fraction of 0.01 of their volume taken by fissures,
   no fissures are present in layer 3;
3. layer 1 has a fraction of 0.01; in layer 2, a fraction of 0.05 of the volume is
   taken by fissures; layer 3 has no fissures;
4. idem, except that the fissure ratio of layer 2 is variable; it is a smoothed
   function of the slope with some added white noise and ranges between 0.03
   and 0.05;
5. layer 1 has a fraction of 0.05, in layer 2 a fraction of 0.01 of the volume is
   taken by fissures; layer 3 has no fissures.

A summary of the different conditions in each model run is given in Table 2.
   The results of the 16 model runs have been used to determine the extent and
the temporal activity of slope instability under the different conditions. From

Table 2. Model run numbers for the fissure scenarios under varying settings for the initial soil moisture conditions ($\theta(i)$) and for the coefficient of performance (C). Mean effective aperture in all model runs set to 0.002 m.

I) $C=0$

Fissure content ($m^3 m^{-3}$)		dry initial conditions, $\theta = \theta_{FC}(i)$		wet initial conditions, $\theta(i) = \frac{1}{2}(\theta_{FC}(i) + \theta_{Max}(i))$
no fissures		Model run	01	02
layer 1 and 2	0.01		03	04
layer 1 layer 2	0.01, 0.05		05	06
layer 1 layer 2	0.01, 0.03-0.05		07	08

II) $C = 0.002^3$

layer 1 and 2	0.01	Model run	09	10
layer 1 layer 2	0.01, 0.05		11	12
layer 1 layer 2	0.01, 0.03–0.05		13	14
layer 1 layer 2	0.05, 0.01		15	16

every run the total number of unstable pixels, the day on which the pixel is initiated for the first time and the return period, defined as the average period between two events of instability, have been obtained. The results have been summarized in Tables 3 and 4.

The cohesion of the first layer and the higher angle of internal friction in the third layer provide enough additional strength to prevent instability. All landslide activity is limited to the second layer of weathered marl. In the 10x10 m DEM of the Barranco de la Coloma 8 out of the 12303 pixels are potentially unstable because the slope angle exceeds the angle of internal friction of 38° of the second layer. When these pixels do not gain any strength from matric suction they will become unstable. For the triggering of any other pixel in the catchment a positive pore pressure at its potential shear surface is needed. For this reason the minimal slope angle, the maximal height of the perched water table $H(i)$ and the maximal water level in the fissure network have been determined for all the unstable pixels in every run to define the threshold of instability. In addition timeseries with the variations of the perched water table $H(i)$ have been reported for each layer at the outlet of the catchment and at a location on the gully–wall. The variations of $H(i)$ in layer 1 to 3 for the scenarios without fissures and with a variable fissure content are given in Figure 4 under dry and wet initial conditions.

From the sensitivity analysis it appears that fissure flow under the applied conditions has primarily a stabilizing effect on the regolith (Table 3). Although

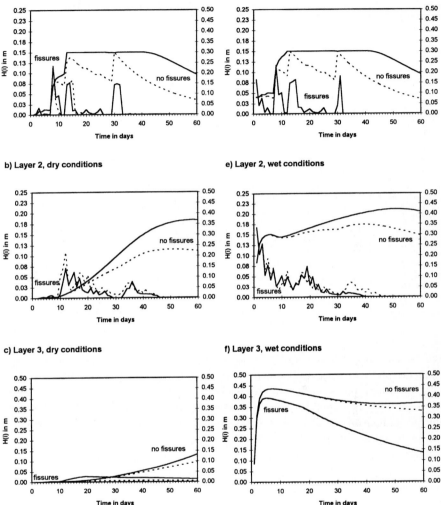

Fig. 4. Comparison of the fluctuations in the height of the perched water table $H(i)$ in layer 1–3 at the outlet of the catchment and at a location on the gully–wall in the period of 60 days (slope angles respectively 1.2 and 31.8°). Graphs are given for the scenarios without fissures and with variable fissure content under dry and wet conditions, a–b–c and d–e–f respectively. Waterheights for the scenarios with fissures are given on the left–hand, the waterheights for the scenarios without fissures are given on the right–hand axis.

Table 3. Summary of the extent of slope instability for the 16 model runs.

Model Run	Number of unstable pixels	Percentage compared to run without fissures	Maximal duration of instability in days
Simulations with dry initial conditions			
01	56	100.0 %	58
03	9	16.1 %	58
05	21	37.5 %	58
07	18	32.1 %	58
09	10	17.9 %	58
11	22	39.3 %	58
13	19	33.9 %	58
15	9	16.1 %	58
Simulations with wet initial conditions			
02	137	100.0 %	60
04	31	22.6 %	60
06	31	22.6 %	60
08	31	22.6 %	60
10	32	23.4 %	60
12	31	22.6 %	60
14	31	22.6 %	60
16	31	22.6 %	60

the safety factor differs slightly for all scenarios due to the varying stress conditions at the shear surface, this can not explain this stabilization. Only in runs 17 and 18, which have a higher fissure content in the first layer than in the second, some stability is gained but even this is less than 1%. Therefore the explanation for the enhanced stability should be sought in the hydrological processes that are included in the model.

For the scenarios without any fissures the total number of unstable pixels is 56 and 137 for the dry and wet initial conditions respectively (run 01 and 02). For all other runs the extent of slope instability decreases. Under wet initial conditions the number of unstable pixels seems to be limited to 32 and does not vary with the fissure content. An explanation of this apparent maximum number lies in the fact that the difference in total matric percolation in a layer and the actual percolation to the next one is used to generate the perched water table $H(i)$. At the beginning of the simulation the percolation of the soil moisture in excess of field capacity leads to a rapid rise in the perched water table and the consequent triggering of the regolith on the steepest slopes. This becomes clear from Table 4 which lists the days on which the unstable pixels are initiated; all instability is triggered in the first 5 days of the timeseries while the important rainfall does not start before day 7. Under dry initial conditions, however, the number of unstable pixels varies with the fissure contents applied (Table 3). For the runs in which no water is routed through the fissure network, i.e. $C = 0$, the

Table 4. Summary of the initiation and the average return periods of slope instability for the 16 model runs.

I) Initiation

Classes (days)	Rain (mm)	Model run															
		01	02	03	04	05	06	07	08	09	10	11	12	13	14	15	16
		(number of unstable pixels per model run initiated within the corresponding period)															
1-5	07	8	95	8	31	8	31	8	31	8	32	8	31	8	31	8	31
6-10	31	1	0	1	0	5	0	3	0	1	0	5	0	3	0	0	0
11-15	32	0	0	0	0	8	0	5	0	0	0	8	0	6	0	1	0
16-20	4	4	5	0	0	0	0	2	0	1	0	1	0	1	0	0	0
21-25	4	8	8	0	0	0	0	0	0	0	0	0	0	1	0	0	0
26-30	30	5	8	0	0	0	0	0	0	0	0	0	0	0	0	0	0
31-35	0	8	13	0	0	0	0	0	0	0	0	0	0	0	0	0	0
36-40	0	11	8	0	0	0	0	0	0	0	0	0	0	0	0	0	0
41-45	0	8	0	0	0	0	0	0	0	0	0	0	0	0	0	0	0
46-50	0	3	0	0	0	0	0	0	0	0	0	0	0	0	0	0	0
51-55	0	0	0	0	0	0	0	0	0	0	0	0	0	0	0	0	0
56-60	0	0	0	0	0	0	0	0	0	0	0	0	0	0	0	0	0
Total:	108	56	137	9	31	21	31	18	31	10	32	22	31	19	31	9	31

II) Return period: absolute frequency distribution

Classes (days)		Model run															
		01	02	03	04	05	06	07	08	09	10	11	12	13	14	15	16
		(number of unstable pixels per model run with average return periods within the limits of the corresponding classes)															
1-<1.5		13	98	9	29	9	25	9	23	9	26	9	21	9	20	9	24
1.5-<2		13	14	0	2	1	6	0	6	0	5	1	4	1	7	0	5
2-<2.5		11	7	0	0	1	0	0	1	0	0	1	2	0	1	0	1
2.5-<3		8	4	0	0	4	0	1	1	0	0	4	2	2	3	0	0
3-<5		9	8	0	0	4	0	1	0	0	0	5	2	0	0	0	1
5-<10		1	4	0	0	2	0	1	0	0	1	2	0	0	0	0	0
10-<20		1	1	0	0	0	0	5	0	1	0	0	0	6	0	0	0
20-<30		0	1	0	0	0	0	0	0	0	0	0	0	1	0	0	0
30-<60		0	0	0	0	0	0	0	0	0	0	0	0	0	0	0	0
≥60		0	0	0	0	0	0	1	0	0	0	0	0	0	0	0	0

number of unstable pixels ranges from 9 to 21 (run 03 and 05) and 18 pixels are triggered in run 07 with variable fissure content. For all runs in which a fraction of the water through the fissure network is routed the number of unstable pixels is increased by one (Table 3). This small increase is consistent with the small change in the height of the perched water table in layer 2 and is limited by the available storage in the fissure network.

In every model run the maximal height of the perched water table and the maximal water level in the fissure network can be combined with the minimal slope angle oberved for the unstable pixels. For the 16 model runs these near–critical pixels with a safety factor close to 1 approximate the threshold of instability of the regolith. From Figure 5 it becomes clear that the generation of positive pore pressures by the perched water table in the second layer is the dominant triggering process. Although the maximal water level in the fissure network reaches the surface in most model runs, the influence of the additional disturbing forces exerted by it is included within the 3.2% of the total of unexplained variance.

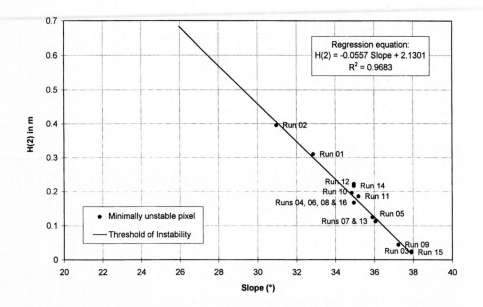

Fig. 5. Threshold of instability: Relation for the unstable pixels in each model run between the observed minimal slope angle and the height $H(2)$ of the perched water table in layer 2; in all scenarios in which fissures are present the fissure network in the pixel is entirely filled.

The 8 potentially unstable pixels of the second layer are triggered in all model runs. Under wet initial conditions instability at these locations is initiated in the first timestep and persists the entire run. Under dry initial conditions they are not triggered before the third day of the timeseries after which they remain unstable as well (Tables 3 & 4.I). The initiation of all other unstable pixels occurs within the wet, first 30 days of the timeseries, the only exceptions being the two model runs in which no fissures are present (run 01 and 02). In these model runs triggering of landslides after the rainfall has ceased can be explained by the slower lateral throughflow in the matrix (Figure 4). As will be expected, the presence of fissures in the regolith seems to speed up both the wetting and drainage of the lower matrix. This can be seen from the distribution of the temporal frequency in Table 4 and Figure 6; a considerable number of the unstable pixels in run 03 to 16 have longer return periods than those in the model runs 01 and 02 which are triggered once and remain active afterwards. So the presence of fissures in the regolith influences in the occurrence of the shallow landslides in two ways. First of all it enhances slope stability as indicated by the decreased number of unstable pixels in comparison with the first model runs without fissures. This is a result of the better drainage by the fissure network through both lateral and vertical losses. Secondly, the fissure network affects the temporal activity of the shallow landslides. The response in matric suction and pore pressure in the second layer is more dynamic as water can drain and infiltrate more freely.

From explorative modelling at different spatial resolutions under different rainfall events and fissure settings it appears that fissure induced infiltration into the lower regolith is highly dependent on the storage capacity available in the fissure network of the second layer. Evidently, if more water can be stored in the fissures, infiltration into the matrix can be maintained over a longer period. This also stresses the influence of the input through the topsoil. If the input of precipitation is diminished or retarded by – for example – the landcover of the upper layer, the replenishment of the fissure network is decreased and the gradient that leads to infiltration into the matrix may not exist. However, when situations exist in which the fissure network is supplied directly through overland flow or through a permeable topsoil, the activity of shallow landslides will be favoured. It can therefore be concluded that the influence of fissure induced infiltration will be relatively high if the storage in the fissure network is large, the routing through the fissure network is low, and it is replenished through a topsoil of high permeability.

Conclusions

The presence of fissures influences slope stability by providing preferential flow paths for infiltration and drainage. Because of the drainage of the saturated out-flow from the matrix, fissures have primarily a stabilizing effect since the extent of landsliding, as measured by the total number of unstable pixels, is reduced for the applied fissure scenarios in comparison with the results for a matrix without any fissures. The storage of water in the fissure network, however, leads to hori-zontal infiltration into the matrix under the resulting gradient. The influence of this by–pass flow on the response time of shallow landslides is reflected in the temporal activity; shallow landslides are initiated earlier and more often in a

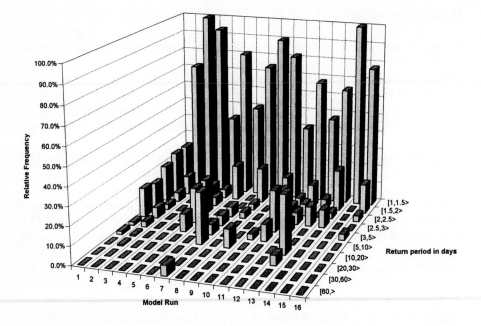

Fig. 6. Relative frequency distribution of average return period of the unstable pixels per model run.

fissured regolith than in a continuous matrix. The importance of fissure–induced, horizontal infiltration into the matrix is dependent on the storage time of water in the fissure network. Factors which increase this storage time are the total storage capacity, limited vertical and lateral losses from the fissure network and a high rate of replenishment, either through a permeable topsoil or directly by runoff into the fissure network.

A concluding remark should be placed on the generation of the perched water tables in the model. The instanteneous generation of the perched water tables under initial wet conditions results not only in the initiation of many landslides at the end of the first timestep, it should also be stressed that it may lead to an underestimation of the importance of fissure–induced infiltration on slope instability; if drainage is not prevented by a corresponding water level in the fissure network, an excessive amount of water from the matrix could be lost directly to the enclosing fissures by saturated, lateral outflow.

Acknowledgments: The authors wish to thank Drs. Derk–Jan Karssenberg for his help with the execution of the model and the Netherlands Geosciences Foundation of the Netherlands Organization for Scientific Research (NWO–GOA) for their financial support.

References

Anderson, M.G. and S. Howes (1985): Development and application of a combined soil water–slope stability model. Q.J.Eng. Geology 18:225–236.

Anderson, M.G., M.J. Kemp, and D.M. Lloyd (1988): Application of soil water finite difference models to slope stability models. In: Bonnard (Ed): Landslides–Proceedings of the Fifth International Symposium on Landslides, Lausanne, 525–530.

Beven, K. and P.F. Germann (1982): Macropores and water flow in soils. Water Resources Research 18:1311–1325.

Bouma, J. (1990): Using morphometric expressions for macropores to improve soil physical analyses of field soils. Geoderma 46:3–11.

Carrara, A., M. Cardinali, F. Guzzetti, and P. Reichenbach (1995): GIS technology in mapping landslide hazard. In: Carrara and Guzzetti (Eds), Geographical Information Systems in Assessing Natural Hazards. Kluwer Academic Publishers, 135–175.

Fredlund, D.G. (1987): Slope stability analysis incorporating the effect of soil suction. In: Anderson and Richards (Eds), Slope Stability. J. Wiley and Sons Ltd., 113–143.

Germann, P.F. (1990): Macropores and hydrological hillslope processes. In: Anderson and Burt (Eds): Process Studies in Hillslope Hydrology. J. Wiley and Sons Ltd., 327–363.

La Roca–Cervigón, N. and A. Calvo–Cases (1988): Slope evolution by mass movements and surface wash (Valls d'Alcoi, Alicante, Spain). Catena Suppl. 12:95–102.

La Roca–Cervigón, N. (1991): Untersuchungen zur räumlichen und zeitlichen Variabilität der Massenbewegungen im Einzugsgebiet der Riu d'Alcoi (Alicante, Ostspanien). Die Erde 122:221–236.

Lane, L.J., M.A. Nearing, J.M. Laflen, G.R. Foster, and M.H. Nichols (1992): Description of the US Department of Agriculture Water Erosion Prediction Project (WEPP) model. In: Parsons and Abrahams (Eds): Overland Flow: Hydraulics and Erosion Mechanisms. UCL Press Ltd. London:377–391.

PCRaster (1996): PCRaster software and manual. For information and downloadable evaluation software, see http://www.frw.ruu.nl/pcraster.html.

Rahardjo, H., D.G. Fredlund, and S.K. Vanapalli (1991): Use of linear and non–linear strength versus matric suction relations in slope stability analysis. In Bell (Ed.): Landslides – Proceedings of the Sixth International Symposium on Landslides, Christchurch, 531–537.

Snow, D.T. (1968): Rock fracture spacings, openings and porosities. J. of Soil Mechanics and Foundations Divisions, Proceedings Am. Soc. of Civil Eng. 94:73–91.

Van Asch, Th.W.J. (1980): Water erosion on slopes and landsliding in a Mediterranean area. Thesis Utrecht University, Nederlandse Geografische Studies, 238 pp.

Van Asch, Th.W.J., and H. van Steijn (1991): Temporal patterns of mass movements in the French Alps. Catena 18:515–527.

Van Steijn, H. (1996): Debris–flow magnitude–frequency relationships for mountaineous regions of Central and Northwest Europe. Geomorphology 15:259–273.

Vauclin, M., R. Haverkamp, and G. Vachaud (1979): Résolution numérique d'une équation de diffusion non lineaire – Application à la infiltration de l'eau dans les sols non–saturés. Presses Universitaires de Grenoble 183.

Wesseling, C.G., D. Karssenberg, W.P.A. van Deursen, and P.A. Burrough (1996): Integrating dynamic environmental models in GIS: the development of a Dynamic Modelling language. Transactions in GIS 1:40–48.

Local Slope Stability Analysis

I. Hattendorf, St. Hergarten, and H. J. Neugebauer

Geodynamics – Physics of the Lithosphere, University of Bonn, Germany

Abstract. Mass movements under the influence of gravity occur as result of diverse disturbing and destabilizing processes, for example of climatic or anthropological origin. The stability of slopes is mainly determined by the geometry of the land–surface and designated slip–horizon. Further contributions are supplied by the pore water pressure, cohesion and friction. All relevant factors have to be integrated in a slope stability model, either by measurements and estimations (like phenomenological laws) or derived from physical equations. As result of stability calculations, it's suitable to introduce an expectation value, the 'factor–of–safety', for the slip–risk. Here, we present a model based on coupled physical equations to simulate hardly measurable phenomenons, like lateral forces and fluid flow. For the displacements of the soil–matrix we use a modified poroelasticity–equation with a Biot–coupling (Biot 1941) for the water pressure. Latter is described by a generalized Boussinesq equation for saturated-unsaturated porous media (Blendinger 1998). One aim of the calculations is to improve the knowledge about stability-distributions and their temporal variations. This requires the introduction of a local factor–of–safety which is the main difference to common stability models with global stability estimations. The reduction of immediate danger is still the emergent task of the most slope and landslide investigations, but this model is also useful with respect to understand the governing processes of landform evolution.

Introduction

The relief of the earth is formed by different tectonic and denudational processes. In the field of landscape evolution description–models and estimations for the quantitative contributions of governing processes are of interest. The explorations consist of observations, measurements and theoretical models. The denudational phenomenons extend over a wide spatial and temporal scale. On the smaller scale erosion and solifluction grind permanently the hills and unplant areas. Even if the soil transfer is hardly visible, landscape–shapes are distinctly formed by those processes. The resulting danger is mostly insidious, but the long–time loss of arable ground and the spreading of deserts can be hazardous.

The danger becomes more evident by the transition to gravitational mass movements, like landslides or mudflows. Generally, they can be subdivided into three classes: *flows*, *slides* and *falls*. The threat with these is more immediate for people and buildings and the emerging costs and risks require a comprehensive analysis of endangered regions. Concerned investigation methods are well established in the field of geotechnical engineering; but until today the knowledge and technique is not sufficient, as you can see in the occurrence of failures even in man–made slopes. Uncontrollable mudflows cause annually several victims as just seen in Italy this year. Such catastrophes are often caused by earth slips in the heights of the mountains. In combination with heavy precipitation the accelerated mass flows irresistibly towards the valleys and buries whole villages under mud and boulder.

Common investigations of endangered slopes include measurements of the shear strength, cohesion and pore water pressure and the prediction of the slip circle. The application of a slope stability model, often based on geometrical and phenomenological considerations (*method of slices*), yields finally a factor–of–safety. This factor describes the ratio between the 'resisting' and 'overturning' forces and serves as a criterion for the appearance of mass movements. But a fundamental problem for the estimations is the punctational character of most received measurements (e.g. drilling). Only the local situation in the ground is explored and mainly scattered information is available. More drill sites enhance the time and cost efforts. Sometimes geophysical measurements are useful to fill the gap (see e.g. Hattendorf and Kümpel 1996). A further problem in common slice–models remains in the lack of knowledge about lateral forces (called *inter-slice forces*). Their stabilizing or destabilizing effects are evident. In this local stability model lateral forces are derived from physical equations. The numerical solutions simulate the displacements and pore water pressure in the slope.

In the following section we will first introduce the generalized Boussinesq equation to describe the soil water flow. After that, the stress in the slope caused by gravity and fluid pressure will be derived from the stationary Navier–Stokes equation with a Biot–coupling. Finally, the principles of local stability analysis and the illustration of the results will finish the article.

The Water Pressure

In saturated–unsaturated porous media the Richards equation (1) is a common description of 3–dimensional fluid flow. It's a composition of the continuity equation and Darcy's law with the peculiarity of a dynamic hydraulic permeability $K(p)$:

$$\phi \, \varrho_f \, g \, \frac{\partial}{\partial t} \Theta(p) = \sum_{i=1}^{3} \frac{\partial}{\partial x_i} K(\Theta(p)) \frac{\partial}{\partial x_i} (p + \varrho_f \, g \, x_3) \tag{1}$$

with the time–variable t, the space–variables $x_{i=1,2,3}$, the water pressure p, the fluid density ϱ_f, the saturation–function $\Theta(p)$ and the porosity ϕ. In the saturated case with a positive fluid pressure ($p > 0$) the permeability becomes constant. In the unsaturated case with negative p values (matric potential) the permeability decreases enormously. This causes a nonlinearity in the equation and increases the requirements for the numerical solution. For our calculations we choose the parameter–functions of vanGenuchten (1980):

$$\Theta(p) = \frac{1}{\left(1 + (\kappa\,|p|)^{\omega+1}\right)^{\frac{\omega}{\omega+1}}} \tag{2}$$

$$K(p) = K_s\,\sqrt{\Theta(p)}\,\left(1 - \left(1 - \Theta(p)^{\frac{\omega+1}{\omega}}\right)^{\frac{\omega}{\omega+1}}\right)^2 \tag{3}$$

with ω and κ as soil–dependent parameters. In Fig. 1 the strong dependence on the water pressure of permeability and saturation is illustrated.

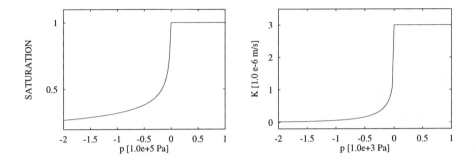

Fig. 1. The saturation function $\Theta(p)$ and hydraulic permeability $K(p)$ for a loamy soil with $K_s = 3 \times 10^{-6}\mathrm{m}^2\mathrm{s}^{-1}$, $\omega = 0.252$, and $\kappa = 8.97$.

First, we imagine an idealized hillslope section like in Fig. 2. The water flow during and immediate after a rainfall event will be mostly vertical, governed by the gravitational field. This process happens in the so–called 'short' time scale. But often, it only takes hours until flows in vertical direction vanishes. The hydrostatic pressure remains and the slower, lateral flows are dominating. Latter process takes place in the 'long' time scale. Blendinger (1996, 1998) showed that for thin domains such a decoupling of the processes in the time scale is also mathematically possible. It's called a Dupuit approximation of equation (1). In conjunction with the decoupling a dimension reduction arises. This is helpful to confine the numerical efforts. With respect to the governing processes for landform evolution (like mass movements) and the interest for the 'long' time scale a Dupuit approximation appears suitable. In Blendinger (1996) you find a detailed exposition of the approximation and the whole mathematical proof, especially questions about the existence and convergence.

The following assumptions for the hill section (Fig. 2) have been taken: an infiltration boundary at the surface $g_2(x_1, x_2)$ is realized by a Neumann border; to avoid mathematical trouble, ponding is not possible, so the slope will never get totally saturated; at the lower surface $g_1(x_1, x_2)$ (the designated slip horizon for the stability analysis) an impermeable layer (homogeneous Neumann border) bounds the domain in the depth. After a scaling of the problem and applying

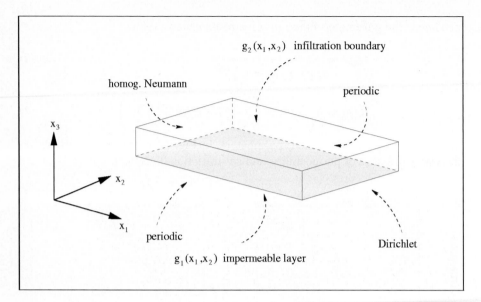

Fig. 2. Idealized hillslope section.

of the perturbation theory, an integration in x_3–direction yields the hydrostatic pressure relation:

$$p(x_1, x_2, x_3, t) = \varrho_f\, g\left[H(x_1, x_2, t) - \left(x_3 - g_1(x_1, x_2) \right) \right] \tag{4}$$

where $\varrho_f\, g\, H(x_1, x_2, t) := p(x_1, x_2, g_1(x_1, x_2), t)$ is defined as the pressure at the lower boundary. We use the abbreviation $\xi := g_2 - g_1$ for the layer thickness. As result of the Dupuit approximation we get the generalized Boussinesq equation for saturated–unsaturated porous media:

$$\phi\frac{\partial}{\partial t} \int\limits_{H}^{H-\xi} \Theta(p)\, \mathrm{d}p - \sum_{i=1}^{2} \frac{\partial}{\partial x_i} \int\limits_{H}^{H-\xi} K(p)\, \mathrm{d}p\, \frac{\partial}{\partial x_i}(H + g_1) = S \tag{5}$$

with S as a source or sink term to enable the infiltration of precipitation water. This nonlinear partial differential equation is of an advection–diffusion type for the variable $H(x_1, x_2, t)$. The solution is realized numerically with a fully implicit finite–element scheme for the spatial directions derived of *Galerkin*'s method. The nonlinearity, affected by the dependence of the hydraulic permeability on the water pressure, is solved by a partly linearisation and a fixed–point iteration.

In Figs. 3 and 4 you see a simulation result of water flow in an artifical hill consisting of loamy soil (parameters according to Fig. 1). For simplicity, the g_1 and g_2 planes are assumed to be parallel. A water divide at the top (Neumann border) and a Dirichlet boundary (e.g. a water reservoir) at the bottom border are assumed. The remaining boundaries on the left and right side are periodic.

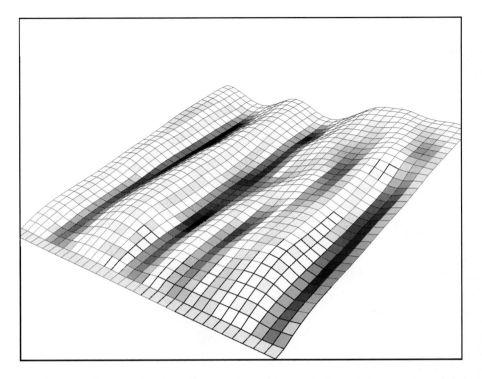

Fig. 3. Water flow on a hillslope (512 m × 512 m) in the stationary state resulting from a continuous precipitation of $400\,\mathrm{mm\,yr^{-1}}$. Dark areas indicate high, light areas low water pressure H.

This test should demonstrate the behaviour of the water flow and the correctness of the numerical solution. In the initial state the region is almost dry and precipitation events increase the water content successively. The fluid will flow downwards in the porous medium driven by the gravitational force. Focused in the valleys, the flux increases. Approaching the Dirichlet boundary, the constant pressure value will suck out the superfluous water. Further precipitation compensate the water loss that the whole system becomes quasi–stationary (Fig. 3). The effect of an enormous precipitation event on the hill slope section is shown in Fig. 4. As assumed, the water pressure has been increased in the 'flooded' domain. We return to this simulation example in the last chapter due to a presentation of the factor–of–safety. Summerized, in this section we have introduced the generalized Boussinesq equation to describe the soil water flow in the slope stability model.

The Stress–Strain Relation

In this section an equation for the displacements of the soil matrix will be derived. The displacements itself are caused by the stress in the porous medium

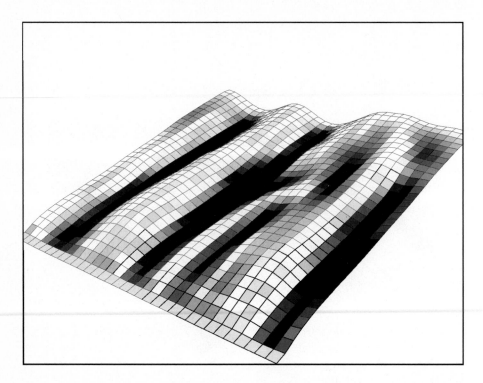

Fig. 4. Water pressure after an extreme precipitation event (150 mm in 3 days), starting from the stationary state shown in Fig. 3. Explanations see Fig. 3.

and therefore mainly decisive for the slope stability. A further aim is to keep numerical efforts and computional system–resources in a reasonable manner. Thus, a dimension–reduction of the 3-D problem would be desirable. If the governing equation could be decoupled analogously to the Richards equation (1), an integration in x_3–direction would achieve the required result.

The most simple approach for the rheology of a porous medium is a linear-elastic assumption. With regard to the diversity of soil types this seems not to be sufficient for mass movements itself with their viscous or viscoelastic behaviour, but in the stationary case of still stable slopes a poroelastic rheology supplies a suitable description (see e.g. Guéguen and Palciauskas 1994 or Ranalli 1987).

Assuming a thin domain in x_3–direction relative to the lateral extensions, like in Fig. 2, the following boundary conditions are realized: the relief of the earth $g_2(x_1, x_2)$ should be a free surface and at the slip plane $g_1(x_1, x_2)$ the displacement u should vanish, as long as no mass movement occurs. The equation of motion (Navier–Stokes equation) for an elastic medium results by equating the divergence of the stress tensor plus gravitational force $(-\tilde{\varrho}\, g\, e_3)$ to the acceleration:

$$\sum_{j=1}^{3} \frac{\partial}{\partial x_j} \left(\sigma_{ij} - \underbrace{\varsigma \delta_{ij} P(p)}_{\text{Biot-coupling}} \right) - \tilde{\varrho} \, g \, \delta_{i3} = \tilde{\varrho} \, \frac{\partial^2 u_i}{\partial t^2} \tag{6}$$

for $i = 1, 2, 3$. The density $\tilde{\varrho}$ is a composition of the soil ϱ_s and the fluid density ϱ_f:

$$\tilde{\varrho} = (1 - \phi) \, \varrho_s + \phi \Theta(p) \, \varrho_f$$

with the porosity ϕ and the saturation–function $\Theta(p)$. We neglect gravitational effects of the air pressure p_a. In the second term of (6) a Biot–coupling between water pressure and soil matrix is realized (Biot 1941). The coupling–constant ς is given by the relation

$$\varsigma = 1 - \frac{K_t}{K_s}$$

with the bulk modulus K of the porous medium (index t) and the solid phase (index s). $P(p)$ agrees in the saturated case with the fluid pressure p, but it has to be modified in the case of unsaturated flow by $\Theta(p)$ (see e.g. Schrefler and Xiaoyong 1993). With respect to the stability model the matric suction ($p < 0$) has a stabilizing effect on the slope. It's proportional to $\Theta(p) \, p$ for $p > -p_a$ where $p_a = 10^5$ Pa is the constant air pressure. For smaller values of p the increase of the effect abates and will be kept constant like shown in Fig. 5. Quantitatively, the considerations can be described by the relation

$$P(p) = \Theta(p) \, p - \left(1 - \Theta(p) \right) p_a$$

For a more detailed discussion of the matric suction and its effects see Brooks et al. (1998).

Regarding momentary stress–strain in the slope, the Navier–Stokes equation becomes stationary and the acceleration term vanishes. Thus, the temporal evolution of the coupled problem is determined by the Boussinesq equation. Then the elasticity equation can be solved within every time–step. Concerning numerical treatment with finite elements derived of *Ritz's* variation method, we focus on the objective function $\mathcal{F} = \int_V \mathcal{L} \, dV$. It coincides in the stationary case with the potential energy and the integration argument $\mathcal{L} = \mathcal{L}(u_i, \frac{\partial u_i}{\partial x_j})$ (for $i, j = 1, 2, 3$) is called the Lagrange–function. The use of variational calculus obtains the Euler–Lagrange equation, for the state which minimizes the potential energy:

$$\frac{\partial \mathcal{L}}{\partial u_i} = \sum_{k=1}^{3} \frac{\partial}{\partial x_k} \frac{\partial \mathcal{L}}{\partial \left(\frac{\partial u_i}{\partial x_k} \right)} \tag{7}$$

It can easily be shown that in our case $\mathcal{F}$ has the form:

$$\mathcal{F} = \int_V \left(\frac{1}{2} \sum_{i,j=1}^{3} \sigma_{ij} \, \varepsilon_{ji} - \varsigma \, P(p) \sum_{i=1}^{3} \varepsilon_{ii} + \tilde{\varrho} \, g \, u_3 \right) dV \tag{8}$$

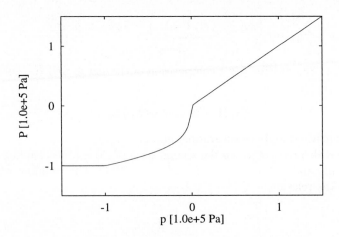

Fig. 5. Dependence of the Biot term P on the fluid pressure p.

If the deformations are sufficiently small, the strain–tensor ε_{ij} and the stress–tensor σ_{ij} are given by following relations:

$$\varepsilon_{ij} = \frac{1}{2}\left(\frac{\partial u_i}{\partial x_j} + \frac{\partial u_j}{\partial x_i}\right) \tag{9}$$

$$\sigma_{ij} = \lambda\,\delta_{ij}\sum_{k=1}^{3}\varepsilon_{kk} + 2\,\mu\,\varepsilon_{ij} \tag{10}$$

with the two material–dependent Lamé–constants λ and μ.

Due to moderate gradients of g_1 and g_2, derivatives of u_3 in lateral directions are small and can be neglected. Hence, the objective function can be minimized with respect to u_3 separately for any given functions of u_1 and u_2:

$$\frac{\partial\mathcal{L}}{\partial u_3} = \frac{\partial}{\partial x_3}\frac{\partial\mathcal{L}}{\partial\left(\frac{\partial u_3}{\partial x_3}\right)}$$

with the result:

$$\bar{\varrho}\,g = \frac{\partial}{\partial x_3}\left\{(\lambda + 2\,\mu)\frac{\partial u_3}{\partial x_3} + \lambda\sum_{i=1}^{2}\frac{\partial u_i}{\partial x_i} - \varsigma\left[\Theta(p)\,(p + p_a) - p_a\right]\right\}$$

The boundary condition of a free surface ($\hat{\sigma}\,\boldsymbol{n}\,(g_2) = 0$ with $\boldsymbol{n} = (-\nabla g_2, 1)$) yields the relation:

$$\sigma_{33}(g_2) = \mu\sum_{i=1}^{2}\frac{\partial u_i}{\partial x_3}\frac{\partial g_2}{\partial x_i}$$

With that we can perform the integration in x_3–direction $\left(\int_{x_3}^{g_2} \mathrm{d}\tilde{x}_3\right)$ and have:

$$\frac{\partial u_3}{\partial x_3} = \frac{1}{\lambda + 2\mu} \left\{ \gamma + \mu \sum_{i=1}^{2} \frac{\partial u_i}{\partial x_3} \frac{\partial g_2}{\partial x_i} - \lambda \sum_{i=1}^{2} \frac{\partial u_i}{\partial x_i} \right\} \tag{11}$$

where

$$\gamma := \varsigma \left[\Theta(p)\,(p + p_a) - \Theta(H - \varrho_f\, g\, \xi)\,(H - \varrho_f\, g\, \xi + p_a) \right]$$

$$+ \phi\, \varrho_f\, g \int\limits_{x_3}^{g_2} \Theta(p) - (1 - \phi)\, \varrho_s\, g\, (g_2 - x_3)$$

and the fluid pressure p given by equation (4):

$$p(x_1, x_2, x_3) = H(x_1, x_2) - \varrho_f\, g\, \big(x_3 - g_1(x_1, x_2)\big)$$

Next, we split the function $\mathcal{F}$ in terms containing displacements of second and first order: $\mathcal{F} = \mathcal{F}_\mathrm{S} + \mathcal{F}_\mathrm{T}$. Concerning the variation, terms independent on $\boldsymbol{u}$ can be neglected. Furthermore, we omit squared terms of the gradient of g_1 or g_2. Thus, for the first term we get:

$$\mathcal{F}_\mathrm{S} = \frac{1}{2} \int\limits_V \sum_{i,j=1}^{2} \left\{ \eta \frac{\partial u_i}{\partial x_i} \frac{\partial u_j}{\partial x_j} + \mu \frac{\partial u_i}{\partial x_j} \left(\frac{\partial u_i}{\partial x_j} + \frac{\partial u_j}{\partial x_i} \right) + \mu\, \delta_{ij} \frac{\partial u_i}{\partial x_3} \frac{\partial u_j}{\partial x_3} \right\} \mathrm{d}V$$

with the abbreviation $\eta := \frac{2\lambda\mu}{\lambda + 2\mu}$, and for the second term:

$$\mathcal{F}_\mathrm{T} = \frac{1}{\lambda + 2\mu} \int\limits_V \sum_{i,j=1}^{2} \left\{ -\varsigma\mu \left[\Theta(p)\,(p + p_a) - p_a \right] \left(2 \frac{\partial u_i}{\partial x_i} + \frac{\partial u_i}{\partial x_3} \frac{\partial g_2}{\partial x_i} \right) \right.$$

$$\left. + \mu\gamma \frac{\partial u_i}{\partial x_3} \frac{\partial g_2}{\partial x_i} + \tilde{\varrho}\, g \int\limits_{g_1}^{x_3} \left(\mu \frac{\partial u_i}{\partial \tilde{x}_3} \frac{\partial g_2}{\partial \tilde{x}_i} - \lambda \frac{\partial u_i}{\partial \tilde{x}_i} \right) \mathrm{d}\tilde{x}_3 \right\} \mathrm{d}V$$

To describe the behaviour of lateral displacements in vertical direction, we assume restrictedly that they alter linearly with x_3:

$$u_{1,2}(x_1, x_2, x_3) = \frac{2}{\xi} \big(x_3 - g_1(x_1, x_2)\big)\, \psi_{1,2}(x_1, x_2) \tag{12}$$

We will call $\psi_i(x_1, x_2)$ a 'modified displacement' which is equal to $u_i(x_1, x_2, x_3 = g_1 + \xi/2)$. This assumption agrees with the situation of a shearable, elastic body fixed on a plane at the lower surface g_1 (Fig. 6). After some transformations and an integration in x_3–direction $\left(\int_{g_1}^{g_2} \mathrm{d}x_3\right)$ the first term of $\mathcal{F}$ reads:

$$\mathcal{F}_\mathrm{S} = \frac{1}{2} \int\limits_\Omega \sum_{i,j=1}^{2} \left\{ \frac{4}{3}\, \xi \left[\eta \frac{\partial \psi_i}{\partial x_i} \frac{\partial \psi_j}{\partial x_j} + \mu \left(\frac{\partial \psi_i}{\partial x_j} \right)^2 + \mu \frac{\partial \psi_i}{\partial x_j} \frac{\partial \psi_j}{\partial x_i} \right] \right.$$

$$\left. + \eta\, A_j\, \psi_j \frac{\partial \psi_i}{\partial x_i} + \mu\, A_i\, \psi_j \frac{\partial \psi_j}{\partial x_i} + \mu\, A_j\, \psi_i \frac{\partial \psi_j}{\partial x_i} + \frac{4}{\xi}\, \mu\, \delta_{ij}\, \psi_i\, \psi_j \right\} \mathrm{d}\Omega$$

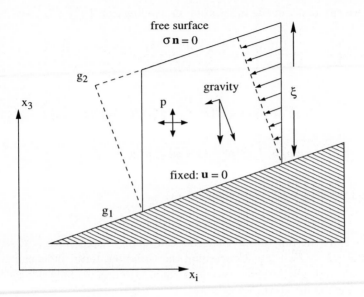

Fig. 6. Fixed and shearable poroelastic body.

with Ω as the remaining 2–dimensional domain and

$$\mathcal{A}_i := -4 \left(\frac{2}{3} \frac{\partial \xi}{\partial x_i} + \frac{\partial g_1}{\partial x_i} \right)$$

The term $\mathcal{F}_T$ includes the saturation function $\Theta(p)$. If we choose equation (2) analogously to the Richards equation, we have to solve the occurring integrations numerically. Using the 'heavy–side' function for Θ as a quite simple approximation, an analytical solution is possible. Finally, $\mathcal{F}_T$ has the form:

$$\mathcal{F}_T = \int_{\Omega} \sum_{i=1}^{2} \left\{ \mathcal{B}(H,\xi) \frac{\partial \psi_i}{\partial x_i} + \mathcal{C}(H,\xi) \frac{\partial g_1}{\partial x_i} \psi_i + \mathcal{D}(H,\xi) \frac{\partial g_2}{\partial x_i} \psi_i \right\} \mathrm{d}\Omega$$

with the pre–factors $\mathcal{B}, \mathcal{C}$ and $\mathcal{D}$ dependent on the water pressure H, layer width ξ and the selected saturation function.

A numerical solution of $\mathcal{F}$ yields the modified displacements ψ_1 and ψ_2. With these and equation (11) the stress–tensor (9) is completely determined. With respect to the stability model we are mainly interested in $\hat{\sigma}$ at the shear surface g_1. Hence, we need the derivatives of $\boldsymbol{u}$ at g_1:

$$\frac{\partial u_i}{\partial x_j}(g_1) = -\frac{2}{\xi} \frac{\partial g_1}{\partial x_j} \psi_i$$

$$\frac{\partial u_i}{\partial x_3}(g_1) = \frac{2}{\xi} \psi_i$$

$$\frac{\partial u_3}{\partial x_3}(g_1) = \frac{1}{\lambda + 2\mu}\left[\gamma(g_1) + \frac{2}{\xi}\psi\,(\lambda\,\nabla g_1 + \mu\,\nabla g_2)\right]$$

for $i, j = 1, 2$. To get explicitly the force $d\boldsymbol{f}$ on an effected area dA, we have to calculate:

$$d\boldsymbol{f} = dA\,(\hat{\boldsymbol{\sigma}}\,\boldsymbol{n})$$

with the normal vector $\boldsymbol{n}$ of dA. In the next section the obtained results of the Boussinesq and the Navier–Stokes equation have to be fitted in the local stability model.

The Stability Model

The introduction of a factor–of–safety F serves as a criterion for the appearance of gravitational mass movements. This means, it should emphasize the stability situation in the ground. The slope stability is mainly governed by forces of geometrical, hydrological or rheological nature. As already mentioned, we are interested in a local factor–of–safety $F = F(x_1, x_2)$ to point out the probability of disturbances for every slope–element. But this requires the hardly measurable lateral forces and the water pressure distribution for every site. In most common models (e.g. Bishop 1955, Morgenstern and Price 1965, 1967, Janbu 1973 or Sarma 1973) the difficulties with unknown forces are avoided by phenomenological considerations and requirements for equilibriums of forces and momentums. The experience often justifies the applications with reasonable and immediate results, but often the single F–value for the whole slope turns out to be disadvantageous. Instabilities within the domain could be overlooked. Local amplifications as well as spreading and alterations of disturbing processes cannot be investigated.

To illustrate the force relations, we imagine a vertical slice along an artificial slope like in Fig. 7. The shown element has the form of a prism with the base dA. A suitable definition for the factor–of–safety is the quotient between the 'actual available' shear strength and the destabilizing momentums, the 'mobilized' shear strength at the slip surface g_1 (e.g. Bromhead 1992). More obvious is the quotient of all resisting (like cohesion and friction) and disturbing (called 'overturning') forces. Both definitions are equivalent.

Here, the stress tensor $\hat{\boldsymbol{\sigma}}$ includes the effects of the lateral forces, the pore water pressure and the gravitational force. First, we expand the problem in normal direction $\boldsymbol{n}$ of g_1:

$$\boldsymbol{n} = \frac{1}{\sqrt{1 + |\nabla g_1|^2}}\begin{pmatrix} -\nabla g_1 \\ 1 \end{pmatrix}$$

The normal stress N along $\boldsymbol{n}$ is then given by the relation:

$$N = (\hat{\boldsymbol{\sigma}}\,\boldsymbol{n})\,\boldsymbol{n}$$

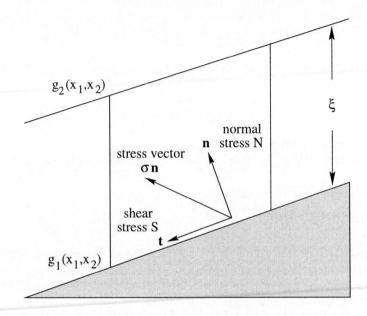

Fig. 7. Slice of a slope element.

The shear stress S acts perpendicularly to n in the slip zone:

$$S = (\hat{\sigma}\, n)\, t$$

where t is a tangential vector ($|t| = 1$ and $t\, n = 0$). To find the smallest factor–of–safety, we have to search for t in direction of the maximum shear stress. Geometrical considerations yield:

$$t_{\max} = \frac{\hat{\sigma}\, n - \left((\hat{\sigma}\, n)\, n\right) n}{\sqrt{(\hat{\sigma}\, n)^2 - \left((\hat{\sigma}\, n)\, n\right)^2}}$$

Further, we need a failure criterion for the appearance of sliding. Useful is the Mohr–Coulomb hypothesis for materials that can't load much shear stress, like i.g. geological substances:

$$S_{\max} = N \tan \vartheta + c$$

with the friction–angle ϑ and the cohesion c, dependent on the material–properties. The equilibrium is valid for still stable slopes and the factor–of–safety is weighting both sides, so we get:

$$F = \frac{S_{\max}}{S} = \frac{N \tan \vartheta + c}{S}$$

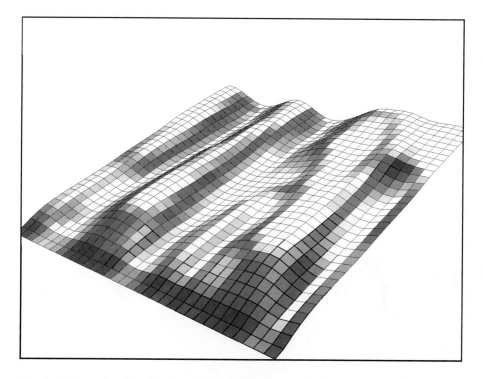

Fig. 8. Factor–of–safety distribution for laterally constant fluid pressure ($H = 0.5\,m$). Dark areas indicate low, light areas indicate high F–values.

Inserting the normal and the maximum shear stress, we finally obtain for the local factor–of–safety:

$$F(x_1, x_2) = \frac{(\hat{\boldsymbol{\sigma}}\,\boldsymbol{n})\,\boldsymbol{n}\,\tan\vartheta + c}{\sqrt{(\hat{\boldsymbol{\sigma}}\,\boldsymbol{n})^2 - \big((\hat{\boldsymbol{\sigma}}\,\boldsymbol{n})\,\boldsymbol{n}\big)^2}} \tag{13}$$

This equation contains the friction angle and cohesion as specific soil parameters. Further, the stress–tensor includes the Lamé and van Genuchten constants, the system density and porosity which have to be determined by measurements. Also, the topographies of $g_1(x_1, x_2)$ and $g_2(x_1, x_2)$ have to be inserted. To get a stability–description of enlarged regions like in common models (global factor–of–safety), equation (13) has to be integrated over the concerned area. But locally confined disturbances are also able to cause large mass movements because of their interactions with the environment. This effects would be omitted with global stability analysis.

To illustrate the application of the model we consider a simple example. We imagine again the hillslope section of Fig. 3; the relief g_2 and the slip horizon g_1 are parallel. The rheological parameters are: $\lambda = 1.73 \times 10^9$ Pa, $\mu = 1.15 \times 10^9$ Pa, $c = 20\,\text{kPa}$, and $\vartheta = 20°$. To emphasize the effects of rheology and topography

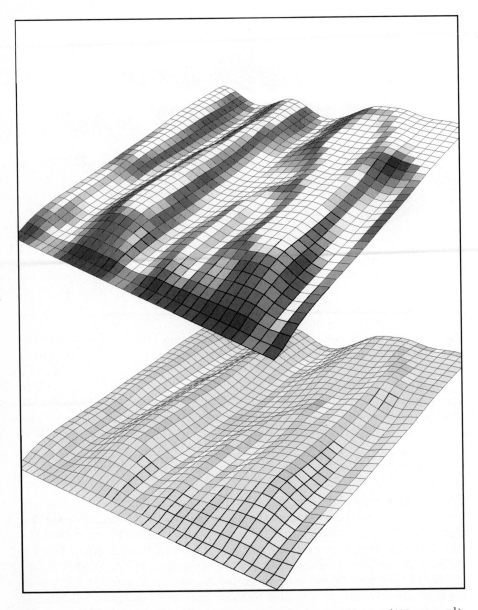

Fig. 9. Above: factor–of–safety distribution under steady precipitation $(400 \, \mathrm{mm \, yr^{-1}})$; below: relative alterations of F compared to Fig. 8: dark decrease and light increase of F.

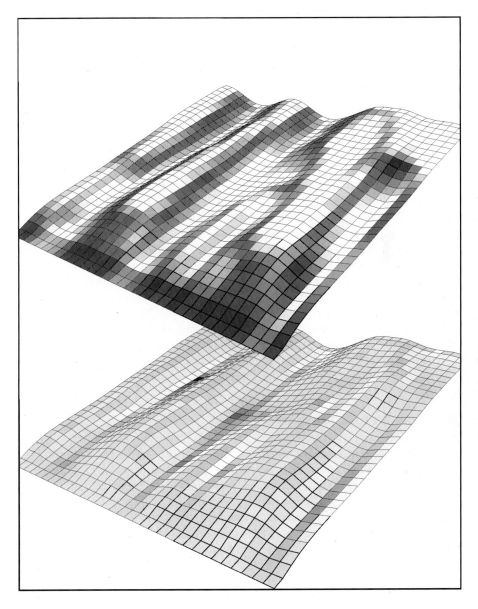

Fig. 10. Above: factor–of–safety distribution after an heavy precipitation event (150 mm in 3 days); below: relative alterations of F compared to Fig. 8: dark decrease and light increase of F

on the slope stability, in the initial state (Fig. 8) we choose a constant water pressure at every point. We see that the distribution of the factor–of–safety is dominated by the geometry. The dark areas provide low F–values which are more instable than the light areas with higher values. This first result demonstrates already the possibility to point out single, endangered slope sections.

Starting the process temporally with precipitation events analogously to the example in the section 'The water pressure' (Fig. 3), the fluid will be focused in the valleys and flow slowly out of the domain at the Dirichlet boundary. The compensation of the water loss by moderate rainfall transfers the system in a quasi–stationary state. According to this water pressure distribution, in Fig. 9 the factor–of–safety is shown. To emphasize alterations of F, we have subtracted the values of the initial state (Fig. 8). Dark regions have decreased and the light regions increased F–values. We see only small variations and conclude that in this case the pressure effect is negligible.

As in the former example the hill section is exposed to an heavy precipitation event (Fig. 4). As result the weight of a slope–element and the pore water pressure grow ('hydrological loading'). The factor–of–safety decreases obviously in regions with high water content and large pressure gradients (Fig. 10). The disturbing forces are spreading and destabilizing the slope mainly in the valleys and rills. For this areas the possibility for slope failures and mass movements is enlarged, but not yet critical because of their small slope–gradients. But if the water flux is hindered to flow from the hills to the valleys, the stability could dangerously decrease. Thus, the importance of drainages is evident and this model enables the simulation of several distinct scenarios. Further, the user could apply different disturbing influences to find out stability alterations and to improve his understanding of the governing processes acting in the slope.

Conclusions and Outlook

In this article a slope stability model with a local factor–of–safety has been presented. The considerations are based on well established physical equations. For the soil water flow we used a generalized Boussinesq equation derived of Richards equation for saturated–unsaturated porous media. A modified stationary Navier–Stokes equation yields the displacements of the poroelastic soil–matrix. The one–sided coupling between pore water pressure and displacements is realized by an additional Biot–term. The aim was to illustrate local stability for every slope element to point out interactions of different forces within the domain. Possible applications are considerable for drainages or larger diggings. Endangered slopes can be systematically analysed by varying the disturbing forces. Even before and after a downcutting of a slope foot the stability alterations can be examined. The calculations of diverse scenarios can optimize remedial and corrective measure to improve the slope stabilization.

However, the local slope stability analysis presented here is just one step towards a physically based risk assessment. The question under which conditions a small local instability is able to induce a large landslide is still open. Hence, coupling of stability analysis and mass movement in order to investigate the feedback of small mass movements on slope stability will be a central point in further research.

Acknowledgment. The authors would like to thank the German Research Foundation (DFG) for financial support within the Collaborative Research Center (SFB) 350.

References

Biot, M.A. (1941): General theory of three dimensional consolidation. J. Appl. Phys. 12: 155–164.

Bishop, A.W. (1955): The use of the slip circle in the stability analysis of earth slopes. Géotechnique, 5: 7–17.

Blendinger, C. (1996): An approximation of saturated–unsaturated Darcy flow in thin domains (in German). University of Bonn, SFB 256, Preprint No. 485.

Blendinger, C. (1998): A Dupuit Approximation for Saturated–Unsaturated Lateral Soil Water Flow. In this volume.

Bromhead, E.N. (1992): The stability of slopes. Chapman & Hall, Blackie Academic & Professional Press, Glasgow, 109–142.

Brooks, S.M., M.G. Anderson, T. Ennion, and P. Wilkinson (1998): Exploring the potential for physically–based models and contemporary slope processes to examine the causes of Holocene mass movement. In this volume.

van Genuchten, M.Th. (1980): Predicting the hydraulic conductivity of unsaturated soils. Proc. Soil Science Soc. of America, Vol. 44, No.5: 892–898.

Guéguen, Y., and V. Palciauskas (1994): Introduction to the physics of rocks. Princeton University Press, Princeton, New Jersey, 135–158.

Hattendorf, I., and H.-J. Kümpel (1996): Investigation of landslides with the electromagnetic induction method (in German). Thesis, Institute of Geology, University of Bonn.

Janbu, N. (1973): Slope stability computations. In: Hirschfeld, E., and S. Poulos (eds.): Embankment Dam Engineering, Casagrande Memorial Volume. John Wiley, New York.

Morgenstern, N.R., and V.E. Price (1965): The analysis of general slip surfaces. Géotechnique, 15: 79–93.

Morgenstern, N.R., and V.E. Price (1967): A numerical method for solving the equations of stability of general slip surfaces. Computer Journal, 9: 388–393.

Ranalli, G. (1987): Rheology of the earth. Deformation and flow processes in geophysics and geodynamics. Allen & Unwin Inc., Boston, 50–113.

Sarma, S.K. (1973): Stability analysis of embankments and slopes. Géotechnique, 23: 423–433.

Schrefler, B.A., and Z. Xiaoyong (1993): A Fully Coupled Model for Water Flow and Airflow in Deformable Porous Media. Water Resources Research, 29: 155–167.

Part III

Modelling Landform Evolution

Landscape Modelling at Regional
to Continental Scales

M. J. Kirkby

School of Geography, University of Leeds, UK

Abstract. Most work on simulating landscape evolution has been focused at scales of about 1 Ha, there are still limitations, particularly in understanding the links between hillslope process rates and climate, soils and channel initiation. However, the need for integration with GCM outputs and with Continental Geosystems now imposes an urgent need for scaling up to Regional and Continental scales. This is reinforced by a need to incorporate estimates of soil erosion and desertification rates into national and supra–national policy. Relevant time–scales range from decadal to geological. Approaches at these regional to continental scales are critical to a fuller collaboration between geomorphologists and others interested in Continental Geosystems.

Two approaches to the problem of scaling up are presented here for discussion. The first (MEDRUSH) is to embed representative hillslope flow strips into sub–catchments within a larger catchment of up to 5,000 km^2. The second is to link one-dimensional models of SVAT type within DEMs at up to global scales (CSEP/SEDWEB). The MEDRUSH model is being developed as part of the EU Desertification Programme (MEDALUS project), primarily for semi–natural vegetation in southern Europe over time spans of up to 100 years. Catchments of up to 2500 km^2 are divided into 50–200 sub–catchments on the basis of flow paths derived from DEMs with a horizontal resolution of 50 m or better. Within each sub–catchment a representative flow strip is selected and Hydrology, Sediment Transport and Vegetation change are simulated in detail for the flow strip, using a 1 hour time step. Changes within each flow strip are transferred back to the appropriate sub–catchment and flows of water and sediment are then routed through the channel network, generating changes in flood plain morphology.

The CSEP/SEDWEB model uses a one–dimensional hydrological model to simulate vegetation growth from monthly climate data. Growth of semi–natural vegetation is simulated using mass balances for living biomass and soil organic matter. Erosion is estimated from the rainfall and runoff distribution, using thresholds related to the vegetation, and used to provide a climatic soil erosion potential (CSEP). Current work is integrating the CSEP with topographic data derived from a global DEM at 1 km horizontal resolution. At this scale, the sediment yield is related to local relief. There is the potential to link the model to both global tectonics, including isostatic components and flexure driven by the erosion, and to orographic rainfall schemes which respond to major changes in topography over geologic time spans.

Simulation models for landscape evolution have been largely based on a knowledge of processes at the scale of a single hillslope profile. Even at this scale, there are at least three substantial challenges:

1. the development of explicit and physically based links between process rates and climate, to allow simulation over the periods required for substantive evolution, estimated as 10^5–10^6 years,
2. understanding and incorporating the dynamic links between slope profile evolution and concurrent changes in soil and vegetation,
3. fully extending our knowledge of slope profiles into three dimensional landforms, with implications for a fuller understanding of drainage density.

A number of important areas of scientific development are driving a need to scale up from small area, short period studies to larger time and space scales. Improvements in both dating and detailed climatic histories is allowing us to re–evaluate the record of Holocene environmental change, with an emphasis on the sedimentary record of medium to large river valleys. There is an urgent need to incorporate estimates of soil erosion and other desertification processes into national and supra–national policy, making use of the best available scenarios for landuse and climate change. In using the output from GCMs and as their resolution improves, we are also moving into a position where we should be more involved in refining the important interactions from topography and surface hydrology to the climate system. On a geological time–scale, it is also increasingly important to recognise the strong interactions between erosion and tectonics which is currently being demonstrated for many passive continental margins (Gilchrist and Summerfield 1994). Relevant time–scales therefore range from decadal to geological, and, in general, there is a clear tendency for problems at coarser scales to be associated with longer time spans (Fig. 1). Detailed studies are typically of three types; detailed process rates and mechanisms, with small spatial and temporal extents; distribution studies, for example based on remote sensing, which give a snapshot of broad spatial extent; or stratigraphic studies which, in principle, provide a time–slice of history at a point. Of course, replicates add to the time dimension for distributions studies and the space dimension of stratigraphic studies, but many of the major problems in the Earth and Environmental Sciences require an extension of both the space and time extent of well founded field studies. For example, an understanding of river catchments in areas of a few square kilometres, should be based on instrumental records for tens to hundreds of years to understand the spectrum of significantly large events. Studies of environmental change typically focus on time periods up to the whole of the Holocene, and commonly on catchments of several hundreds to thousands of square kilometres (e.g. Macklin et al. 1992, Bull 1991). Studies of plate tectonics are at least continental in scale and relate to periods of millions to hundreds of millions of years. Other areas in Fig. 1 can be the basis of valid studies, but present severe difficulties in interpretation. For example we may construct a model for the evolution of a hillslope profile over millions of years, but this is only valid if we make restrictive assumptions about the tectonic history, divide stability, climatic independence and other factors which depend on a broader spatial context. Similarly, in looking at global change, we are wise to confine ourselves to a particular historical setting, with its implications for a particular sequence of weather and tectonic events.

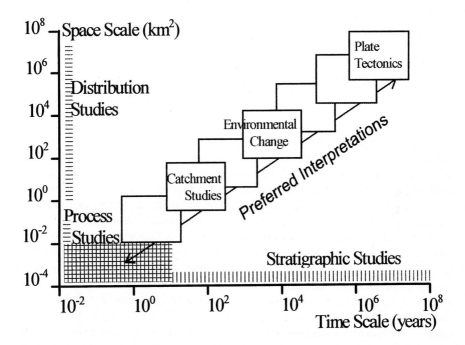

Fig. 1. Space and time scales for Geo–system research (modified from Kirkby and Cox (1995), Fig. 1).

Approaches to Large–Area Modelling

A shared understanding at regional to global scales is a critical part of any fuller collaboration amongst environmental scientists working on Continental Geo–systems, and there are clear needs to use numerical modelling to extrapolate from the domains of detailed knowledge to the study of coarse scale and long period processes and forms. Three methods of approaching an understanding at coarse scales are

1. create models which are directly applicable at a coarse scale,
2. rely on self similarity to bridge between scales, and
3. nest models at different scales.

There are clear advantages in addressing the coarse scales directly, both in the relative simplicity of suitable models and in their direct relevance to geophysical problems. At this scale erosion models are typically based on diffusion, at rates which are constant or spatially varying, although the best current models (e.g. Howard et al. 1994, Kooi and Beaumont 1996)) incorporate a much more realistic representation of fluvial processes in particular. Diffusion models can be shown to be asymptotically appropriate to represent a wide range of water–driven processes (Kirkby 1971), but are not appropriate when landforms are far from 'mature', as is commonly the case in actively tectonic, and therefore many geomorphically interesting situations. There is at least a risk, therefore, that the models may not be readily related to a recognisable process reality.

An assumption of self–similarity bypasses many of the scale issues and implicitly assumes that at least some processes behave as power laws across a range of scales. However many studies indicate that simple scaling remains valid across only a limited range, and that there are characteristic scales at which the relationships break down (Hovius et al. 1997), for example at the area above which clear channels first appear in the landscape, and therefore related to drainage density. However, there has been considerable success in explaining some aspects of landscape geometry, provided that these thresholds are not crossed (e.g. Stark 1991, Turcotte 1994). The main difficulty with this approach, although one that is likely to be overcome progressively, is the difficulty in explicitly linking self similarity results with what we have learned from process studies.

The third approach, and the one which most explicitly links process studies with coarser scale models is by attempting to nest models within one another, and so build from scale to scale. In principle there must be a formal integration over space and/or time or other form of reconciliation between one scale and the next. Such a reconciliation is rarely trivial, so that progress on this path has been slow. Its potential appears to be great, however, and this approach emulates the ways in which scale problems are dealt with in other physical sciences; through a mixture of aggregation or disaggregation, combined with a number of principles and laws which have direct relevance at a range of scales. Nesting implies that small areas are combined into larger areas, so that the changes within a larger area can only be approximately re–constituted by a statistical disaggregation. This provides a simpler model, but with some inevitable loss of resolution.

Two approaches to the problem of scaling up are presented here for discussion. The first (MEDRUSH) is to nest representative hillslope flow strips into sub–catchments which sub–divide a larger catchment of up to 5,000 km^2, creating an explicit two–stage nested model structure. The second is to link one–dimensional models of SVAT type within DEMs at up to global scales (CSEP). This may be seen as primarily a large scale approach, although it will be shown that this approach can be reconcile with an understanding of small scale processes.

The MEDRUSH Model

The MEDRUSH model (Kirkby et al. 1995) is being developed as part of the EU Desertification Programme (MEDALUS project), primarily for semi–natural vegetation in southern Europe over time spans of up to 100 years. Catchments of up to 2500 km^2, such as the Agri (Basilicata) or Guadalentin (Andalucia), are divided into up to 200 sub–catchments (Fig. 2) on the basis of flow paths derived from DEMs with a horizontal resolution of 50m or better. Within each sub–catchment a representative flow strip is selected, strongly biased towards the major flow axis of the sub–catchment. Hydrology, sediment transport and vegetation change are simulated in detail, separately for each representative flow strip, using a 1 hour time step, and the resulting changes in surface roughness, stoniness, soil properties and channel density influence flow strip response in subsequent time steps, with strong interactions between each of the components (Fig. 3). This work is based on a detailed flow strip model previously developed (Kirkby et al 1996). Interactions between the variables along the flow strip take place at a range of time scales. For example, sediment responds to surface hydrology instantly, and selective transportation progressively deposits fines or creates an armour layer. Vegetation develops in response to soil moisture through an ecologically based sub–model, over a period of weeks. Erosion roughens the surface by cutting rills, and deposition smooths it, while rainsplash tends to fill depressions and the growth of vegetation mounds slowly roughens it. Erosion of the soil gradually changes its properties, usually making it more stony in shallow Mediterranean soils. All of these changes produce significant feed–back to the hydrology and erosion processes over the periods of decades which are significant for the evolution of abandoned land, although their effects are usually masked where the land is cultivated.

Hydrology, sediment yield and long term changes within each flow strip are transferred back to the appropriate sub–catchment, assuming similar changes within each elevation band (for that sub–catchment). Flows of water and sediment are then routed through the channel network which links sub–catchments, generating changes in flood plain morphology which are based on the generated flood frequency distribution.

MEDRUSH is clearly a physically based model, which is explicitly nesting sub–catchments within the whole catchment, and representative slope profiles within each sub–catchment. In any such nested approach, it is important not to claim too much detail as the model output is disaggregated. Since each sub–catchment is represented by a single representative flow strip, the distribution of change within the sub–catchment is a statistical description, and can only be diagrammatically mapped back to the map of the sub–catchment. Such a loss of detail is inevitable in a nested model, and detail can only be partially regenerated by mapping on the basis of elevation bands or other correlated variables, rather than directly, point by point.

The output from each sub–catchment is combined into a flow–routing scheme for water and sediment along the main channel network. The whole model is coded in C++ and closely integrated within the GRASS Geographical Information System (Fig. 4), which provides a shell for input of data and visualisation of the output. Validation is provided partly by discharge records for the whole catchment, but we are also developing fuzzy validation methods which take note of qualitative characteristics of the sub–catchments, such as erosion intensity

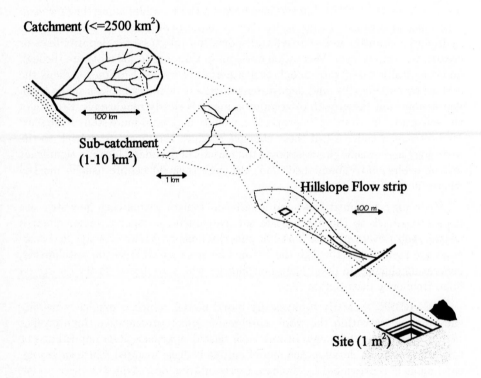

Catchment (<=2500 km²)

Sub-catchment
(1-10 km²)

Hillslope Flow strip

Site (1 m²)

Fig. 2. Nesting of scales in the MEDRUSH model.

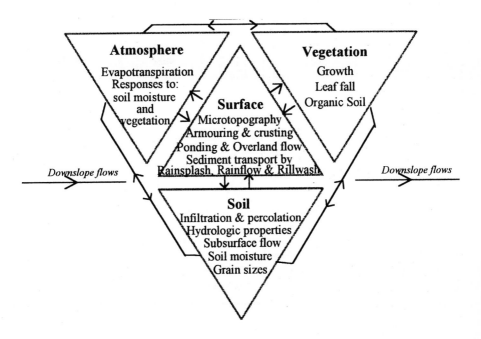

Fig. 3. Interactions between atmosphere, soil, surface and vegetation in the MEDALUS flow strip model.

and surface appearance (mainly for abandoned land). To assess the impact of global change, climate generators based on GCM output from the Hadley Centre model are being combined with simple alternative land use modifications to provide scenarios for future decades.

One Dimensional Regional Models

At regional to continental scales, one effective modelling strategy is to combine a one dimensional (vertical) model based on water balances and hydrological processes in the vertical with an implicit or explicit lateral transfer scheme. The lateral transfer is inevitably based on a mass balance, which maybe applied for both water and sediment, and is valid in an appropriate form at all relevant scales from local transfers along the hillslope to large catchment scales. This mass balance takes the form, for a raster array:

$$\text{input} - \text{output} = \text{net increase in storage}$$

in general,

$$\frac{\partial z}{\partial t} + \nabla \cdot S = a \qquad \text{with} \qquad \nabla \cdot S = \frac{\partial S_x}{\partial x} + \frac{\partial S_y}{\partial y}$$

where

$$z = \text{elevation (for sediment) or water depth (for overland flow)},$$

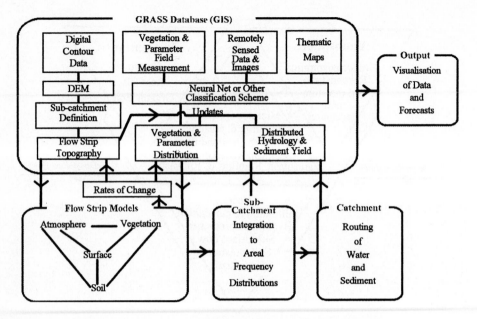

Fig. 4. Integration of MEDRUSH model with GRASS GIS.

t = time,

S = water or sediment flux, and

a = net external accumulation rate (by net rainfall, dust etc.).

Where sediment flux is less than the transporting capacity and is partially limited by supply, as commonly occurs for fine sediments, it maybe necessary to supplement the mass balance with a second equation, for the sedimentation balance:

$$\text{sediment pick–up} = \text{detachment} - \text{deposition}$$

in one dimension,

$$\frac{dS}{dx} = (U - \frac{S}{h})$$

where

U = unconsolidated detachment rate,

x = horizontal distance,

h = mean sediment travel distance, and

α = ratio of consolidated to unconsolidated detachment (< 1).

Removal and transport tends to be

'supply limited' at distances $\gg \dfrac{h}{\alpha}$

'flux limited' at distances $\ll \dfrac{h}{\alpha}$, for which sediment transport is close to its transporting capacity, $U \cdot h$.

As the area of interest becomes larger, the proportion of the area dominated by lateral transfers increases, but the absolute size of the unit areas within which one–dimensional vertical transfers are a good approximation also increases. Table 1 below suggests the orders of magnitude of relevant scales for 1–D and lateral transfers.

Table 1. Unit areas over which 1–D schemes are thought to be valid.

Size of region of interest (km^2)	Name of region	Domain dominated by 1–D transfers (km^2)	% of Area	Name of 1–D domain
10^{-6}	Site	10^{-6}	100 %	Site
10^{-2}	Hillslope	10^{-3}	10 %	Plot
10^{2}	Catchment	10^{0}	1 %	Sub–catchment
10^{6}	Continental	10^{3}	0.1 %	GCM sub–grid

The one–dimensional vertical models are relatively robust to scale changes, but the role of lateral transfers is more closely linked to the scale of interest. At local scales the lateral fluxes are more intimately and interactively linked to the vertical fluxes, for example through hydraulic and surface gradients, while at coarse scales the main links are through river valleys and flood plain sediment storage. It is therefore argued that, for regional or continental scales, the only important lateral transfers are essentially those along major rivers, and that lateral transfers at smaller scales can, to a reasonable order of approximation, be ignored. Thus the lateral transfers are, at coarser scales, increasingly decoupled from the vertical transfers except over very long time scales ($\gg 10^4$ years), and the vertical balances can be used to model vegetation, water balance, and sediment yield directly.

The one dimensional vertical models are closely allied to SVAT (Soil Vegetation Atmosphere Transfer) schemes in GCM surface interactions, but are being proposed here as a means of modelling regional changes at the surface. Clearly there are implications for improving SVAT schemes in GCMs, but these are not pursued here. 1–D schemes are generally built on a water balance, which can then be used to drive a vegetation growth or equilibrium vegetation model. Other models are built on top of this basic scheme.

A simple example of a 1–D model is the Climatic Peat Potential (CPP) model for the growth of ombrogenous peat mires (Kirkby, Kneale, Skellern and Lewis 1995). In this case, potential peat sites are wet enough to evapotranspire at potential rates year–round, so that the water balance is simplified. Seasonal drying of the surface layers, based on mean monthly climate data, defines the surface zone (acrotelm) in which aerobic decomposition of the organic matter takes place. Peat can only accumulate where the addition of organic matter exceeds the total decomposition rate, so that the potential for peat growth is largely dependent on a climatic balance based on a 1–D model (Fig. 5). This model can be used to understand global peat distributions, but can also be used to model the growth of individual bogs by taking account of lateral flows within the peat, driven by peat depth and gradient. For an individual bog, the condition for peat growth is that the CPP>0. For a blanket bog (hill peat), the equilibrium depth is proportional to CPP multiplied by the Topographic Wetness Index (the ratio of drainage area per unit contour length divided by local gradient). For a raised bog, the maximum depth is proportional to $(CPP)^{0.5}$ multiplied by bog diameter.

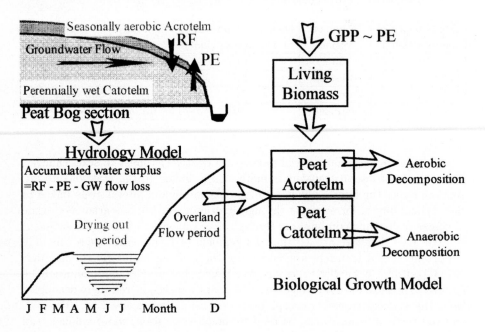

Fig. 5. Climatic Peat Potential Model.

A second model based on a 1–D water balance is the CSEP, or Cumulative Soil Erosion Potential (Kirkby and Neale 1987, de Ploey et al. 1991, Kirkby and Cox 1995). The mean monthly climate data used for the CSEP model should include some data on rainfall intensity, usually described by the number of rain days or distribution of daily rain amounts for each month. For a wider range of climates, actual evapotranspiration is constrained by available water to grow biomass in proportion. Living Biomass and organic soil are modelled separately, and both contribute to the threshold for overland flow runoff, which is compared with the frequency distribution of daily rainstorms (Fig. 6). The CSEP model simulates vegetation growth from monthly climate data, currently using IIASA (Leeman and Cramer 1991) interpolated data for a $(\frac{1}{2})° \times (\frac{1}{2})°$ latitude–longitude grid. Semi–natural vegetation is 'grown' either to equilibrium or in transient mode using mass balances for living biomass and soil organic matter. Runoff and (Runoff)^2 are integrated over time, using thresholds related to the vegetation, and the latter is used to provide a climatic soil erosion potential (CSEP). Current work is integrating the CSEP with topographic data derived from a global DEM at 1 km horizontal resolution.

This model is consistent with both simple local process models and event–based period erosion models. The formal reconciliation across space and time scales is important, not only to show consistency with more detailed approaches, such as MEDRUSH, but also to understand and estimate the responses to changing conditions. On a spatial scale, starting from the scale of the slope profile, the rate of sediment transport at each point by soil erosion processes at the slope base may be estimated from many sources (summarised in Carson and Kirkby, 1972) as taking a form similar to:

$$S \propto \Psi x^2 \Lambda \approx \Psi H x$$

where

$$\Psi = \text{CSEP},$$
$$x = \text{horizontal distance downslope},$$
$$\Lambda = \text{slope gradient, and}$$
$$H = \text{slope relief.}$$

Thus on a regional scale, the specific sediment yield from the slopes is estimated as proportional to ΨH. At this scale, the sediment yield is therefore related to local relief, which is derived from local differences in the elevation of neighbouring points, taken from a 1 km or 10 km resolution DEM.

On a temporal scale, the expression for sediment transport above may be integrated from the form for a single storm which is taken as:

$$S(r) \propto q^2 \Lambda = (p(r-h)x)^2 \Lambda \qquad \text{for} \qquad r \geq h$$

where

$$S(r) = \text{sediment transport from a storm of rainfall } r,$$
$$q = \text{total overland flow discharge per unit width from the storm},$$
$$h = \text{threshold rainfall before runoff begins, and}$$
$$p = \text{proportion of runoff once it begins.}$$

Summing over the frequency distribution of storms (on a monthly basis),

$$S = \sum_r N(r)\, S(r)$$

where

$$S = \text{total average annual sediment transport and}$$
$$N(r) = \text{number of days per year with rainfall } r.$$

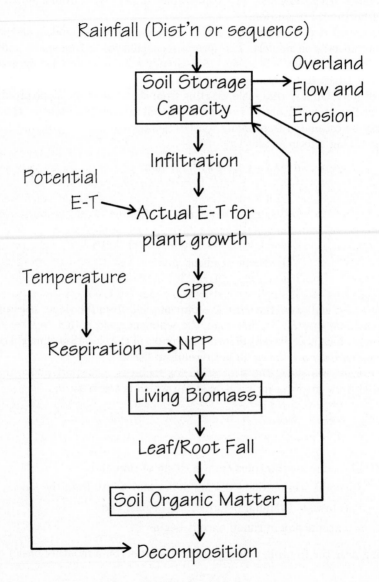

Fig. 6. Flow chart for CSEP model.

This sum can be made explicit for particular frequency distributions, of which the simplest is a simple exponential or $\Gamma(1)$ distribution. For this case the total sediment transport may be integrated as:

$$S = p\,R\,r_0\,\exp(-h/r_0)\,x^2\,\Lambda$$

where

$$R = \text{total (monthly or annual) rainfall and}$$
$$r_0 = \text{mean rain per rain--day.}$$

This expression should be evaluated from monthly values for rainfall (R and r_0) and storage capacity (h). More complex distributions may be used, including other Gamma distributions or the sum of two simple exponentials, but it is clear that the parameters of the daily rainfall distribution appear explicitly in the annual sediment yield estimate. Annual cycles of vegetation also have a strong and explicit impact, through changes in the storage capacity h.

The CSEP model, although primarily a lumped 1–D model, can therefore be rationally reconciled with the temporal distribution of storm intensities, the annual cycles of change in vegetation cover (cultivated or semi–natural), and with the spatial detail of sub–grid topography, through the relief. It is also hoped to incorporate global scale lithology into the final estimate. At the next stage, the SEDWEB model, built upon the CSEP, is intended to route local sediment yields through major drainage systems to provide oceanic sediment and solute yields, related to past and present conditions. In the longer term, there is the potential to link SEDWEB to both global tectonics, including isostatic components and flexure driven by the erosion, and to orographic rainfall schemes which respond to major changes in topography over geologic time spans.

Over decadal time spans, the CSEP potential vegetation is also being compared with remotely sensed land cover classes to provide measures of regional degradation and sensitivity to global change, creating a Regional Degradation Index (RDI) which is also being developed within the MEDALUS III project. The CSEP allows natural vegetation to grow in response to the climate, but can also be adapted to grow particular crops, or allow various degrees of modification of the natural vegetation by grazing or gathering of plant materials. The potential cover is being compared with a historic (15 year) record of actual cover interpreted from remotely sensed imagery, to provide estimates of relative erosion potential from actual and potential cover (Fig. 7). This provides a tool to estimate the degree of desertification by erosion and degradation of vegetation cover which can be applied to large areas, using a unified methodology, and also allows estimates to be made of sensitivity to prospective changes in climate or landuse.

Conclusion

There are still substantial difficulties in applying our knowledge of detailed processes at coarser space and time scales than the Hectare–Year. Nevertheless, a

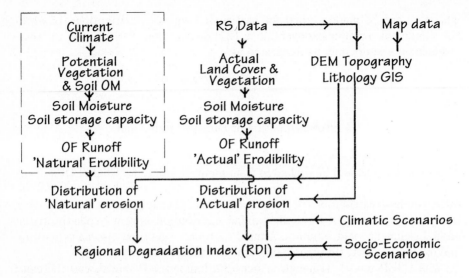

Fig. 7. The Regional Degradation Index (RDI), using the CSEP model to compare 'natural' and 'actual' erosion estimates.

number of methods are emerging which attempt to face these problems systematically and provide avenues for further development. Although some problems may be solved, in principle, by the use of increasingly powerful computers, it is critical to retain a clear comprehension of the significant problems at each scale of interest, and ensure that the most significant interactions at each scale are observed in the field and incorporated into the associated models. There is no unique path forwards, but several avenues show that there is a significant future for physically based models at coarse as well as fine scales.

Acknowledgement: Much of the work described in this paper has been supported through the EC, DG XII in the MEDALUS projects (I, 1991–92: II, 1993–95: III, 1996–98).

References

Bull, W.B. (1991): Geomorphic responses to Climatic Change. Oxford, 326pp.

Gilchrist, A.R., and M.A. Summerfield (1994): Tectonic models of passive margin evolution and their implications for theories of long–term landscape development. In Process models and theoretical geomorphology, ed M.J. Kirkby. John Wiley, 55–83.

Hovius, N., C.P. Stark, and P. A. Allen (1997): Sediment flux from a mountain belt derived by landslide mapping. Geology 25(3): 231–34.

Howard, A.D., W.E. Dietrich, and M.A. Seidl (1994): Modelling fluvial erosion on regional to continental scales. J. Geophysical Research, 99(B7), 13: 971–86.

Kirkby, M.J. (1971): Hillslope process–response models based on the continuity equation, Transactions. Institute of British Geographers, Special Publication 3: 15–30.

Kirkby, M.J., A.J. Baird, S.M. Diamond, J.G. Lockwood, M.D. McMahon, P.L. Mitchell, J. Shao, J.E. Sheehy, J.B. Thornes, and F.I. Woodward (1996): The MEDALUS slope catena model: a physically based process model for hydrology, ecology and land degradation interactions in Mediterranean Desertification and Land Use (ed. Thornes J.B. and J. Brandt), John Wiley, Chichester, 303–354.

Kirkby, M.J., and N.J. Cox (1995): A climatic index for soil erosion potential (CSEP) including seasonal and vegetation factors. Catena 25, 333–52.

Kirkby, M.J., P.E. Kneale, S.L. Lewis, and R.T. Smith (1995): Modelling the form and distribution of peat mires, in Hydrology and Hydrochemistry of British Wetlands, (ed. Hughes J.M./R. and A.L. Heathwaite). John Wiley, 83–93.

Kirkby, M.J., M.L. McMahon, and R. J. Abrahart (1995): MEDALUS II Final Report (EC DG XII).

Kirkby, M.J., and R.H. Neale (1987): A soil erosion model incorporating seasonal factors. International Geomorphology, II, John Wiley, 289–210.

Kooi, H., and C. Beaumont (1996): Large scale geomorphology – classical concepts reconciled and integrated with contemprary ideas via a surface processes model. J. Geophysical Research–Solid Earth, 101(B2): 3361–86.

Leeman R., and W.P. Cramer (1991): IAASA data base for mean monthly values of temperature, precipitation and cloudiness on a global terrestrial grid. Report RR–91–18, IAASA, Laxenburg, Austria, 62pp.

Macklin, M.G., B.T. Rumsby, and T. Heap (1992): Flood alluviation and entrenchment: Holocene valley floor development and transformation in the British Uplands. Geological Society of America, Bulletin 104: 631–43.

Ploey, J. de, M.J. Kirkby, and F. Ahnert (1991): Hillslope erosion by rainstorms – a magnitude–frequency analysis. Earth Surface Processes & Landforms 16: 399–409.

Stark, C.P. (1991): An invasion percolation model of drainage network evolution. Nature, 352, (6334): 423–25.

Turcotte, D.L. (1994): Modeling Geomorphic processes. Physica D, 77(1–3): 229–37.

Exploring the Potential for Physically–Based Models and Contemporary Slope Processes to Examine the Causes of Holocene Mass Movement

S. M. Brooks, M. G. Anderson, T. Ennion, and P. Wilkinson

School of Geographical Sciences, University of Bristol, England

Abstract. Recent research has advanced the application of physically–based soil hydrology–slope stability models to elucidate Holocene slope instability processes. Since investigations of past mass movement present particular problems for model parameterisation and validation, contemporary slope failures need to be utilised. One field site, Glen Feshie, has provided model parameterisation data which has enabled investigation of temporal changes in the probability of slope instability resulting from progressive soil development. A second field site, Glen Livet, has allowed some model validation and closer investigation of the hydrological processes which govern slope instability. A third field area on the Mendip Plateau has provided more evidence that slope failure in temperate regions might well be governed by negative rather than by positive pore water pressures. From these investigations, greater emphasis needs to be placed on unsaturated zone hydrology and slope stability. A major issue for future model development involves the inclusion of hydrological and geotechnical processes associated with developing vegetation. Since this largely involves the unsaturated zone of the soil profile, it is suggested that future model development and application needs to concentrate more on detailed and accurate process representation in 1–dimension to overcome some of the inadequacies which currently persist in interpretations of Holocene slope stability. Hence important issues can be highlighted which need to be retained in new models which can handle larger areas and include long time intervals, suitable for elucidating Holocene slope stability.

Introduction

Recent use of physically–based models to elucidate shallow translational slope failure occurring in the Holocene has highlighted several important issues which provide essential guidance for future research. The four main issues which have emerged centre around; firstly, modelling the role of hydrological behaviour in evolving soil profiles, with emphasis on the unsaturated zone (Anderson et al. 1987, Brooks and Anderson 1995); secondly, fulfilling parameterisation requirements for increasingly sophisticated physically–based models; thirdly, validating such models where applications relate to previous time periods (Konikow and

Bredehoeft 1992) and; finally, development of physically–based models to give closer consideration to the operation of processes internal to the system which, to date, have been inadequately represented (Grayson et al. 1992). Modelling Holocene slope stability poses particular problems related to parameterisation and validation since the occurrence of slope failure, essential for model validation, precludes opportunities for model parameterisation. Conversely, the existence of conditions appropriate for geometric, geotechnical and hydrological parameterisation is only possible for slopes which have not failed. Hence validation opportunities are not provided. This paper discusses such issues in relation to contemporary slope processes and their potential to combine with the development of physically–based models to elucidate Holocene slope failure.

A central conclusion emerging from recent research into contemporary slope processes involves the significance of soil profile hydrology for slope stability. Research from the 1970's had assumed that soil profile hydrology was only significant in as far as it facilitated the development of fully saturated "worst case" conditions (Carson and Petley 1970, Carson 1975). Under the assumption that all slopes have experienced such "worst case" conditions at some point in their history, their maximum stable slope angle is equal to the semi–frictional value if there is no cohesion. Thus, the geotechnical properties of the soil were emphasised in explanations of changing long–term slope instability. If such conclusions are sustainable, then the development of physically–based models could disregard the role of soil hydrology and be further developed to include better representation of changing geotechnical behaviour. However, more recent field and modelling results have challenged these initial conclusions. From field evidence, Anderson et al. (1980) found that thresholds for slope instability are equally explicable in terms of partial saturation of the soil profile, and that different soil profiles provide varying opportunities for water–table movement which is instrumental in dictating the maximum stable slope angle. This field analysis considered only scenarios where the water–table could attain a "worst case" position between the ground level and the shear surface. Since then, field and modelling analyses have shown that under certain conditions, the water–table will not even rise above the shear surface and that unsaturated conditions prevail for all storms (Anderson et al. 1987, Brooks et al. 1993, Brooks and Anderson 1995). In these circumstances unsaturated zone hydrology is critical to slope stability as it dictates the minimum negative pore water pressure (or suction) that can develop at the shear surface under different storms. While such conclusions were originally formulated for tropical regions having deep, permeable soils (Anderson et al. 1988), further assessment for temperate regions has provided similar conclusions for both coarse–grained soil profiles and for profiles which are relatively poorly developed. In general, soils of high permeability are the ones commonly involved. Study of contemporary slope processes has thereby provided a clear indication that physically–based models should focus on soil profile hydrology, as well as on geotechnical behaviour, particularly if they are to be applied to investigations of past mass movements.

A second major issue relates to model parameterisation requirements and, in particular, to the need for increasingly large data sets to suit the demands of state–of–the–art models. Through the 1990's there has been considerable emphasis on developing models which can handle large areas, while retaining a high degree of spatial and temporal resolution. Concurrent model development has also involved up–grading geometrical representations through the number of dimensions included. 1–dimensional schemes have been overtaken by 2–, 3– and even 4–dimensional models. This produces huge demands on parameterisation data which are difficult to meet. The problem is compounded when model applications relate to past periods, as is the case for recent applications of physically–based models to Holocene slope stability (Brooks et al 1995a, 1995b, Brooks and Collison 1996). Given the problems of obtaining opportunities for both parameterisation and validation where previous periods are being considered, careful consideration needs to be given to choice of model scale and dimension. This paper will show how points raised in relation to processes governing slope stability, in particular unsaturated zone hydrology, provide rather different priorities for future model development. Highly developed 2– or 3–dimensional schemes may not necessarily provide the best opportunity for investigating Holocene slope instability.

The third issue of model validation has seen some very interesting discussion in recent publications (Konikow and Bredehoeft 1992, Beven and Binley 1992, Fawcett et al. 1995, DeRoo 1996), which has considerable bearing on model applications to the Holocene. Traditionally, close matching of field measurement with model prediction was viewed as the most persuasive evidence that model performance was adequate. The progression to larger, more sophisticated models challenges this traditional emphasis for validation. The issue raised by Beven (1989), and emphasised in Grayson et al. (1992), that point measurements from the field cannot be used to validate models, becomes increasingly relevant as model scale is increased, and discretised spatial and temporal units become larger. A possible compromise involving internal validation, suggested by Fawcett et al. (1995), offers potential for reliable validation schemes appropriate for modelling long–term changes in slope stability. Internal validation is based around comparison of field and model results at an intermediate stage. Given that it is the interplay between temporal trends in soil hydrological behaviour and climatic change that is central to long–term changes in slope stability, validation centred on hydrological output is justified, regardless of whether the slope actually fails. By adopting this approach possibilities for both parameterisation and for validation are presented.

Fourthly, the case for using physically–based models for research lies not so much in their predictive capabilities, but in their ability to elucidate the operation of complex processes and the way these may vary through time. The issue of process representation is therefore worthy of close consideration (Grayson et al. 1992). This paper will focus on this in particular, with respect to unsaturated zone hydrological and geotechnical behaviour for vegetated slopes. While recent research has correctly emphasised the development of models which represent unsaturated zone hydrology in detail, the development of linked routines to represent the role of vegetation has been limited (e. g., Brooks et al. 1995b). The role of changes in vegetation cover have been discussed widely in relation to Holocene slope stability (Innes 1983, Brazier and Ballantyne 1989), but process representation for vegetation in physically–based models has lagged behind the representation of soil hydrological processes. Specifically, better routines are required for both the hydrological (Running and O'Loughlin 1988, Feddes et al. 1988) and the geotechnical (Waldron et al. 1983, Sidle 1991, 1992, Wu and Sidle 1995) roles of vegetation, since both assume varying significance to slope stability depending on the depth of the shear surface in relation to rooting patterns (Greenway 1987). Preliminary simulations reported in Brooks and Collison (1996) indicate that vegetation has a net stabilising role if the shear surface is located within the rooting zone, attributable to root reinforcement and to large negative pore water pressures compared with unvegetated slopes. Finally, in this respect we need to be aware that many slope stability models treat landslides as rotational and apply the Bishop method for stability analysis. In many instances, failure is non–circular and models need to take this into account by adopting different stability analysis methods, especially where vegetation creates a hydrological and geotechnical discontinuity which is planar.

This paper, therefore, utilises a coupled soil hydrology–slope stability model, which is undergoing continual development (see Anderson et al. 1996), to try to elucidate the causes of mass movements during past periods in the Holocene. Contemporary slope failures are used to provide a basis for model development, parameterisation and validation. It will be shown how results from earlier research (Brooks et al. 1993, 1995a, 1995b, Brooks and Richards 1994) have demonstrated the potential for models to elucidate changes in slope stability which may have characterised certain slopes developing through the Holocene. However, results of recent laboratory experimentation and further modelling of contemporary slope failures have indicated that future model applications must address the important issues of parameterisation requirements and process representation since, in certain circumstances, these factors have important implications for interpretations of past slope failure.

Process Representation in the Coupled Soil Hydrology–Slope Stability Model

The coupled soil hydrology–slope stability model recently used in the elucidation of Holocene mass movement occurrence has been developed over the past decade. Currently, $2\frac{1}{2}$–, 2– and 1–dimensional hydrological schemes are available, while slope stability assessment is possible using either 1– or 2–dimensions. Hitherto, only the 1–dimensional scheme has been applied to investigate changing soil hydrology and slope stability in the Holocene for three reasons. Firstly, there are considerable parameterisation requirements of the $2\frac{1}{2}$– and 2–dimensional schemes. Secondly, detailed representation of the soil profiles using small cells, each having different hydrological and geotechnical properties, has only been possible in 1–dimension until recently and, thirdly, process representation where soils remain unsaturated is adequately modelled in 1–dimension since flow is vertical in the freely–draining soils considered. The fact that soils in slope foot locations are often found to be wetter than those upslope is usually attributable to flow convergence. The implication is that this involves a decreasing slope gradient (i. e., concave slope) which may not represent the most likely location of shallow translational failure. The interplay between slope gradient change, hydrological response and mass movement requires further research using 2– or 3–dimensional models. For the purposes of this paper, limited advantage is gained by adopting $2\frac{1}{2}$– or 2–dimensional representation relative to the cost of providing adequate parameterisation for these larger model schemes.

The 1–dimensional scheme presently available is shown in Fig. 1. It consists of a vegetation sub–routine which treats the canopy as a store from which some of the rain is lost via evaporation and also through stemflow. If the storage capacity is exceeded rain is no longer intercepted and reaches the ground surface directly. The vegetation routine also allows for enhanced permeability in the surface soil layer (the depth of which can be specified) as well as enhanced mechanical reinforcement in the rooting zone (Collison 1995). Evaporation can take place from the soil surface, following a diurnal variation described by a sinusoidal function with time as the independent variable.

Recent model applications have highlighted the importance of negative pore water pressures (suctions) in controlling slope stability (Brooks and Anderson 1995). Suctions on the potential shear surface provide additional strength to the soil, and to ascertain the "worst case" condition it is important to consider the lowest suction achieved during storms. In the current scheme the omission of transpiration uptake processes might be critical to establishing representative "worst case" conditions. Ongoing developments are addressing this problem, but the applications described below involve shear surfaces found to be at greater depths than the rooting zone. Furthermore, the effect of changing root density with depth needs to be included. Hence initial investigations of the combined effects of vegetation, climate and soil development on slope stability in the Holocene have been possible.

Hydrological representation involves solution to the Richards equation, with flux obeying D'arcy's Law. Within the unsaturated zone hydraulic conductivity varies with moisture content and pore water pressure (negative), while in the saturated zone it is constant. The Millington and Quirk (1959) routine is used to describe the suction–moisture content–hydraulic conductivity relationship, although other routines are now available (Campbell 1974, Van Genuchten 1980). Hysteresis in the soil moisture characteristic curve is not included since slope failure largely takes place during wetting cycles in shallow soils. Rainfall is applied over a specified iteration period, fluxes are calculated for each interval, and water content–hydraulic conductivity–pore water pressure is updated. At the end of each hour the pore water pressures for each cell are used in an infinite planar slope stability analysis to find the factor of safety variation with depth and time.

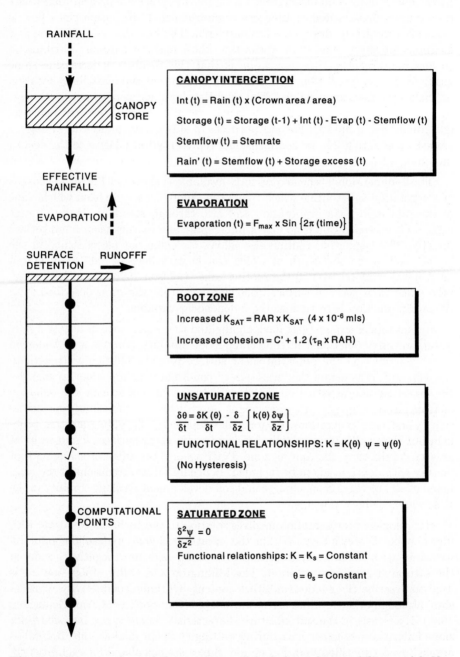

Fig. 1. 1–dimensional coupled vegetation–soil hydrology–slope stability model. New applications to 2–D (hydrology/stability) and $2\frac{1}{2}$–D (hydrology).

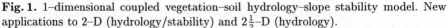

An important facility of the model relates to its treatment of spatial variation in model parameters, an issue commonly ignored in model development schemes and applications. With growing emphasis on the importance of model uncertainty (Binley et al. 1991, Beven and Binley 1992), it is desirable to include assessment of the implications of spatial variability. For each parameter a distribution is defined by its mean and standard deviation. A value for each parameter is randomly drawn from the distribution (although each value does not have an equal chance of being selected) and a number of simulations is carried out, each one involving a different parameter set. Over this series of simulations a factor of safety distribution is derived, reflecting the combined degree of variability in each of the parameters. With high variability in parameters then the factor of safety will also exhibit high variability. However, certain parameters might be associated with high natural variability, but if they are comparatively uninfluential on the factor of safety then the output distributions will show small variability. Conversely, parameters of low variability but high influence may cause factors of safety distributions to be highly dispersed. Clearly, less effort needs to be expended in defining variability for parameters in the former category. While such an approach involving inclusion of parameter variability enables investigation of likely failure probabilities, rather than simply providing a statement of whether the slope is stable or unstable, such analysis can confuse attempts to establish the spatial location of slope failure. This has important implications for model validation, which frequently relies on unique model outcomes for comparison with field data.

Early Model Applications

The earliest application of the 1–dimensional version of the model to investigating slope failure in the Holocene involved simple parameterisation for freely–draining podsols developing on river terrace deposits in Scotland. Model discretisation was also straightforward with division of soil profiles into cells being carried out on the basis of existing horizons. Fig. 2 shows discretisation for the actual soil profiles. Additional cells were actually employed for the unweathered parent material which, for the profiles considered, consisted of freely–draining river gravels with high frictional strength. This initial application highlighted the problem that where simple parameterisation is possible for past scenarios (using ergodic substitution), validation opportunities are not provided. However, some important conclusions were reached regarding likely changing factor of safety distributions over timescales equivalent to that of the Holocene.

The main changes observed in the soil profiles as a result of translocation and weathering were three–fold. Firstly, the overall depth of the soil above the unweathered parent material showed a progressive increase, from just 25 cm in recently developing soil profiles (@80 years) to around 90 cm in the oldest profiles (10–13 000 years). Secondly, the number of horizons increased, they became better–expressed over the 13 000 year sequence and, importantly, they became increasingly differentiated. Thirdly, each horizon had undergone temporal changes

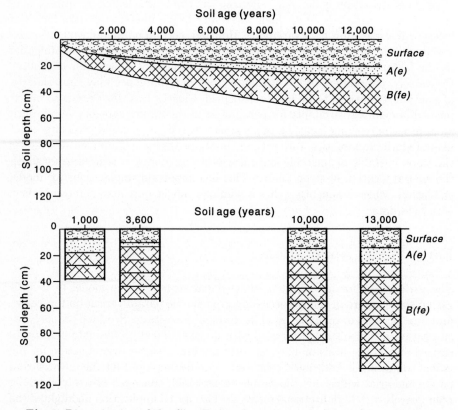

Fig. 2. Discretisation of the Glen Feshie chronosequence for model application.

in porosity, textural, and organic composition, each property being significant to the hydrological behaviour of the soil profile. Each developing soil profile exhibits a unique set of depth, morphological and compositional properties, giving each distinctive geotechnical and hydrological behaviour. Over time the net result involves an increase in the likelihood of shallow translational slope failure.

By running the model with variability included temporal changes and depth variations in factors of safety have been ascertained (Brooks et al. 1993). These are shown in Fig. 3. The evolving hydrological and geotechnical behaviour appears to result in progressive increases in the probability of shallow slope failure, with profiles having greater development being more likely to fail. Fig. 3 emphasises the point that alternative interpretations are possible once the effect of spatial variability is included. This relates both to the depth at which failure takes place and to the stage of development in the soil profiles most conducive to slope failure. There is a clear suggestion that it is only at depths of between 50 and 90 cm that failure will take place, and that this is only likely in the Late Holocene.

The results of such analysis have implications for traditional interpretations of the causes of Holocene slope failure. Secular climatic change has frequently been identified as a major cause of Holocene mass movement in high latitudes (Grove 1972, Starkel 1966, Innes 1983, Brazier and Ballantyne 1989). Some simple scenarios considered using the model, involving 10–day and 5–day periods of zero rainfall, followed by 10–hour storms with rainfall intensities of either $1 \, \mathrm{mm \, h^{-1}}$ or $5 \, \mathrm{mm \, h^{-1}}$, help to establish the relative significance to slope failure of climate, set in the context of progressively developing soil profiles. For each storm a progressive decline in the factor of safety is seen with increasing soil development, but for failure in mature podsols a climatic threshold of 5 rain-free days followed by a rainfall intensity of $5 \, \mathrm{mm \, h^{-1}}$ is required. An intensity of $1 \, \mathrm{mm \, h^{-1}}$ is insufficient to promote failure in any of the soils considered. Ongoing research is addressing this issue of climatic thresholds using 2–dimensional modelling and a wider range of soil profiles.

A further major conclusion of this early research is that for the soil and climate combinations considered, the water–table does not need to rise to the ground surface for failure to take place, as suggested by Carson and Petley (1970). Furthermore, the shear surface remains unsaturated throughout the simulations. The model results thereby provide evidence that temperate zone slope stability can be suction–controlled. This is attributable to the high permeability of the soils considered. Similar conclusions emerging from research into tropical cut slopes (Anderson et al. 1987), were explained by a combination of high permeability and excessive depth of weathered granites. In the tropics, higher rainfall totals and intensities mean that soils need to be deep and permeable for suction to be maintained and thereby be significant to "worst case" scenarios. The lower rainfall totals and intensities of the temperate zone mean that shallow soils may have their stability controlled by negative pore water pressures, but a wider range of permeabilities, involving lower values, can be significant.

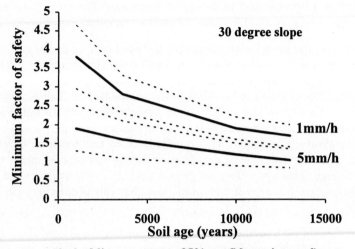

(dashed lines represent 95% confidence interval)

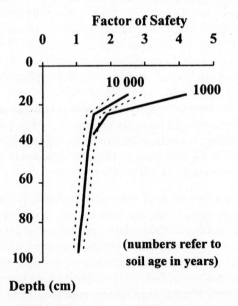

Fig. 3. Temporal trends in Holocene slope stability, including parameter uncertainty for (a) 30 degree slope and 2 rainfall events of 10 hours duration (b) 30 degree slope and 1 rainfall event of 10 hours duration.

In terms of model choice, where unsaturated conditions dominate, flow is vertical and a 1–dimensional representation is appropriate. Given the need for detailed representation of the soil profile and the demands this produces for parameterisation (as well as computational demands to ensure model stability with small cells), 1–dimensional modelling has distinct advantages over $2\frac{1}{2}$– and 2–dimensional schemes. While the 1–dimensional model can handle the development of unsaturated conditions it is clearly an inappropriate choice for a full examination of slope hydrology and stability where saturation is widespread and occurs frequently. In such circumstances, lateral throughflow is important and this cannot be accommodated in a 1–dimensional scheme. It is also important to consider the nature of the failure. Within the stability analysis sub–routine, the 1–dimensional version of the model can only be used for assessment of infinite planar shear surfaces (Nash 1987). Again, for shallow soils with distinct planar failure surfaces, 1–dimensional assessment of stability is appropriate. For circular (rotational) or irregular failure surfaces, more sophisticated forms of stability analysis are required (e. g., Janbu 1954, Bishop 1955), for which 2–dimensional modelling is necessary.

The modelling described above uses contemporary field assessment for model parameterisation, and the results then provide insight into possible causes of slope failure. They indicate to a certain extent, factors which might be important to include in larger scale models for long simulation runs where detailed process inclusion is prohibitive. To date, traditional explanations for late Holocene slope instability have emphasised climatic change. The modelling results permit closer consideration of other factors, such as pedogenesis, and establish whether or not such additional factors warrant further study. The above scenarios proved impossible to validate in the absence of actual slope instability, although limited internal validation might have been possible through field instrumentation of pore water pressures under different rainstorms. However, wider exploration of the region surrounding the field site presented validation opportunities based on actual occurrence of slope instability. This was pursued with the dual aim of exploring issues related to model validation, as well as trying to establish the significance of suction–controlled slope stability in humid temperate environments.

An Opportunity for Validation Using Contemporary Slope Failures

The river terraces used to parameterise the model described above, are located in a region of eastern Scotland which has been glaciated in the Pleistocene. The last ice retreat occurred around 10 000 BP (Young 1971, Sissons 1979), but certain areas not affected by the Loch Lomond Ice Readvance have been ice–free since 13 000 BP. Considering the soil profiles of the river terraces, we would expect small differences between these different–aged older profiles as compared with the greater differences in the younger profiles. This is consistent with general models of soil development, where change is rapid in the first few thousand years and then slows through time (Birkeland 1975, Bockheim 1980).

The soils characteristic of the region as a whole, are podsols similar to those of the older terraces, having well–developed eluvial and illuvial horizons, depths of about 100 cm, little organic matter in any but the surface horizon, relatively sandy texture and relatively low bulk densities. In several locations they are undergoing shallow translational failure, with depths to the shear surface of between 70 and 100 cm. Several such failures are known to have occurred early in 1988, and detailed rainfall data are available for the period.

Model parameter data were obtained from soil profiles located immediately upslope of the failure. These were found to be similar to the parameter sets of the more mature soils described above, which had not undergone failure. One–dimensional modelling was again employed to investigate short–term temporal variations in pore water pressures and factors of safety for the depth corresponding to the shear surface. The results of the modelling show that the factor of safety just reaches unity during a storm which occurred on 9–10th February, 1988. Interestingly, rainfall data indicate maximum hourly intensities of just under $5\,\text{mm}\,\text{h}^{-1}$, with rainfall intensities of $2.5\,\text{mm}\,\text{h}^{-1}$ being sustained for 10 hours. Storms occurring earlier in the winter season, which did not produce factors of safety as low as unity, show that recovery of pore water pressures to their pre–storm levels takes around 4–6 days. Such rainfall conditions, which actually occurred in the region, closely match the climatic thresholds for failure in well–developed podsols found from the simulations described earlier in the paper. Both the model simulations and the actual slopes involve gradients of 30 degrees. In terms of ascertaining appropriate climatic conditions which promote failure in well–developed podsols the model is consistent with field evidence. There is also close agreement concerning the depth to the shear surface, in both cases being around 80 cm.

Results from the hydrological model indicate that the pore water pressures at the shear surface remain negative throughout the rainstorms, even when the factor of safety reaches unity. Again this is consistent with the conclusions reached in the preceding section. Further analysis of slope stability was undertaken using the resistance envelope technique, described in Janbu (1954) and Anderson et al (1987), given the significance of this finding to geomorphological interpretations and to further model development. This technique involves simplifying the slope geometry such that pairs of mobilising shear stresses and normal force combinations can be ascertained for potential failure surfaces at different depths (Fig. 4). By plotting the Mohr–Coulomb shear resistance relationship assuming zero pore water pressures (i. e., shear surface is coincident with water–table), the pore water pressure required to enable the shear strength to equal the downslope shear stress can be found. At certain depths negative pore water pressures are necessary to maintain stability, and these relate largely to the shallower depths in the soil profiles. In the Glen Livet example (Fig. 4b), a suction of around 6 kPa needs to be maintained for stability. This is consistent with results of modelling, where the storms which produced failure in February 1989 involved attainment of suctions of under 6 kPa. In terms of the Holocene, it is these shallow failures which are more likely to dominate, as soils develop from very shallow coverings to deeper profiles. Also, within freely–draining parent materials failure is promoted by the downwards movement of a wetting front rather than by a rise in the groundwater, as envisaged by Carson and Petley (1970) under their "worst–case" scenario. Thus we need to be able to develop and apply hydrological models to slope stability under unsaturated conditions. In such cases, extension to 2–dimensional analysis is of lesser importance than is further consideration of the way negative pore water pressures are generated. Central to this is further development of model sub–routines for better inclusion of factors which relate to unsaturated zone hydrological and geotechnical behaviour.

New Ideas: The Importance of Suction–Controlled Slope Stability

The foregoing analysis has been based on two separate, but related, field sites. One site is strong in providing parameterisation data for modelling the effect of long–term soil development. The other site provides opportunities for model validation using contemporary slope failure. In both situations, it is clear that for freely–draining parent materials, supporting soil profiles developing through the Holocene, unsaturated shear strength is a major factor to consider. Since this has hitherto attracted limited interest from geomorphologists concerned with slope stability, we need to develop better understanding of the range of circumstances under which it is important and, hence, when 1–dimensional hydrological modelling could be used in preference to 2– or $2\frac{1}{2}$–dimensions. In such cases model development can focus closely on specification of process equations which describe unsaturated zone hydrological and geotechnical behaviour.

The central control on whether soils remain unsaturated concerns the relationship between rainfall intensity and water flux through the soil profile (Freeze 1987). In establishing rainfall intensity–duration combinations likely to have prevailed at different times in the Holocene, Brooks et al. (1993) used contemporary storms from Aviemore to group them by generating mechanism. Significant differences were found between cyclonic and anticyclonic storms, which provided a data base for considering the effect of Holocene climatic change on slope stability (Fig. 5a). Typical intensities for anticyclonic storms were around $5\,\mathrm{mm\,h^{-1}}$, while those of cyclonic storms were around $1\,\mathrm{mm\,h^{-1}}$. Allowing for the fact that cyclonic storms tend to be of greater duration, soil profiles were found to be more likely to be unstable under the intensity–duration combination characterising anticyclonic storms.

Similar analysis was performed using rainfall data from the Mendip Plateau in Southwest Britain, where significant differences were again found between storms of different synoptic origin (Fig. 5b). Typical intensities for anticyclonic storms are 5–$6\,\mathrm{mm\,h^{-1}}$, with durations of 5 or 6 hours. For cyclonic storms intensities are commonly around 1–$2\,\mathrm{mm\,h^{-1}}$ and durations are 10–20 hours. Shorter duration frontal storms, of around 10 hours, can produce greater intensities, around $4\,\mathrm{mm\,h^{-1}}$. A high–magnitude rainfall event occurring during 1968 provided further information on the short–term intensity–duration rainstorm char-

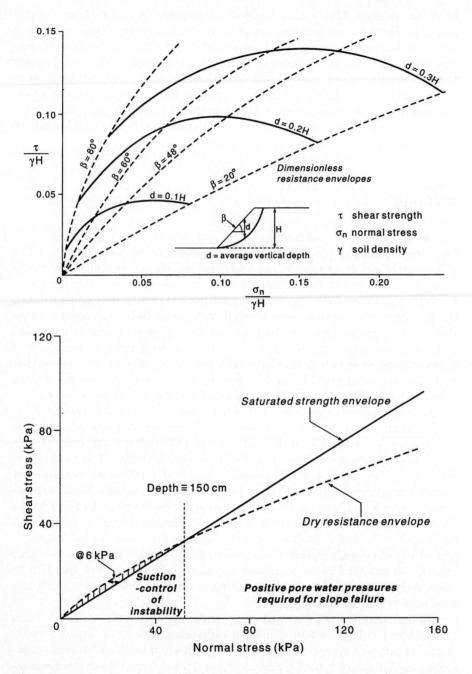

Fig. 4. Resistance envelope method for slope stability assessment. a) Generalised scheme for varying slope geometry and gradient (after Anderson et al. 1987); b) Specific solution for Glen Livet soils and slopes (after Brooks and Anderson 1995).

a) Aviemore

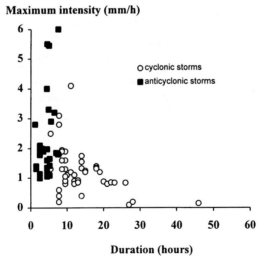

Fig. 5. Intensity–duration relationships for storms grouped by synoptic origin. a) Storms from Aviemore (NE Britain); b) Storms from Mendip Plateau (SW Britain).

acteristics for the Mendip Plateau (Priddy and Blackdown). If we also consider
the classification of storms adopted by the Meteorological Office, "violent showers" have intensities greater than $50 \, \mathrm{mm \, h^{-1}}$, sustained for 0.8 hours, "heavy
showers" require at least $10 \, \mathrm{mm \, h^{-1}}$ over for 3 hours, while "moderate showers"
need to attain $2 \, \mathrm{mm \, h^{-1}}$ for a period of 5 hours. In Fig. 6, these characteristics are compared with the threshold intensity–duration combinations associated
with different types of landsliding (Caine 1981). Under such rainfall, slope failure
is clearly to be expected. Furthermore, such climatic information provides insight into the "worst case" storms likely to be experienced and, when considered
in association with soil types in the region, permits initial investigation of the
likelihood of soil profile saturation.

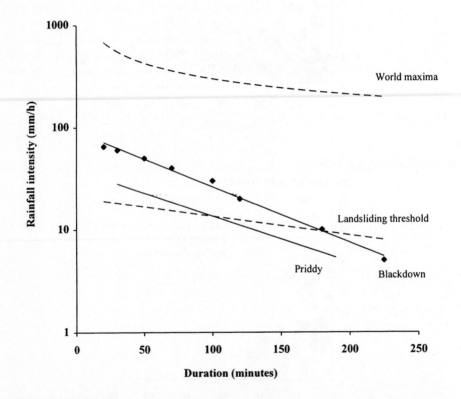

Fig. 6. Rainfall intensity–duration thresholds associated with slope failure (after Caine
1981), showing relationships between rainfall intensity and duration for 2 sites on
Mendip (Blackdown and Priddy).

Soil profiles typical of the Mendip Plateau were considered in association with
the climatic data to investigate the potential for the maintenance of negative
pore water pressures under the extreme rainstorms described above. Typically,
soil profiles of the locality are developed in a loess covering and consist of well–

developed surface and textural B horizons (with significant vertical translocation producing a finer particle size distribution in the B horizon), with a less clearly expressed thinner, coarser, eluvial horizon. The loess is underlain by well–jointed Carboniferous Limestone which is relatively permeable, but situated some 5m below the ground surface. Thus the bedrock is likely to exert relatively little influence on hydrological response in the surface region of the soil. A 2m core was taken from a slope in the locality and instrumented in the laboratory using tensiometers. Rainfall intensities of $60\,\mathrm{mm\,h^{-1}}$, $30\,\mathrm{mm\,h^{-1}}$ and $2\,\mathrm{mm\,h^{-1}}$ were applied, corresponding to the "violent", "heavy" and "moderate" showers defined in the Meteorological Office Classification. Antecedent conditions were varied from 3 to 10 days of drainage preceding the onset of rainfall.

Fig. 7 presents results for pore pressure responses for the surface and B horizons. Under the highest intensity, complete profile saturation was achieved after 1 hour 50 minutes, although the B horizon became saturated after 30 minutes. For the intermediate intensity, 8 hours was required for complete saturation. The A horizon started to respond the rainfall after 30 minutes. Finally, under the lowest intensity it was 50 hours before the profile became completely saturated and the surface horizon only showed an initial response after 6 hours. For the highest intensity, with just 3 days of drainage producing initial suctions of $145\,\mathrm{kPa}$, violent showers can produce saturated conditions after 30 minutes in the B horizon. However, for the majority of rainfall intensity–duration combinations likely to be experienced as shown in Fig. 6, these soil profiles will remain unsaturated. Suction seems to dominate slope instability, even for rainstorms which are likely to exceed the landsliding thresholds identified by Caine (1981).

Model Development for Analysis of Holocene Slope Stability under Unsaturated Conditions

The above discussion has highlighted the potential for using physically–based models for elucidating Holocene mass movement processes. In addition to issues concerning model parameterisation and validation, it is critical to use results from modelling exercises to question process representation and the implications this has for the conclusions reached. The hydrological subroutines use the solution of the Richards equation including D'arcy's Law and the principle of continuity. This enables matric flux to be calculated which, in the unsaturated zone, operates vertically. It does not take into account the effect of macro–pore by–passing flow, which could be important in certain circumstances. In particular, macropore flow can be responsible for rapid groundwater recharge and if the groundwater system is located at depth, then macropore flow could enable high suctions to be maintained within the soil. In the examples described above, the soils are relatively coarse–grained (sands in the Scottish example and silts in the Mendip region), so macropore flow is of lesser importance than matric flow. To extend such an approach to finer–grained soils would require reformulation of the hydrological routine to include macropore flow. Also, when dealing with soil profile saturation the scale of analysis needs to be extended to 2–dimensions.

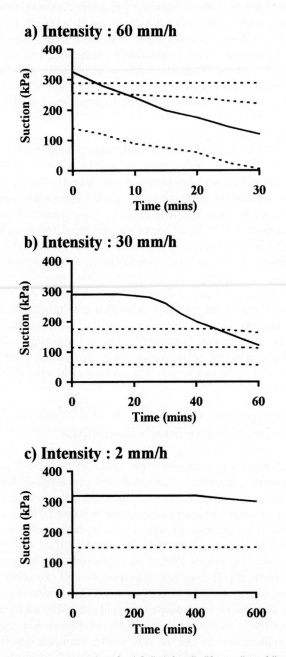

Fig. 7. Pore water pressures associated with "violent", "heavy" and "moderate" showers (solid line represents A horizon; dashed line represents B horizon with varying initial suctions).

Since the hydrological representation seems appropriate for the examples considered above, attention needs to be paid to the stability analysis. Shallow translational failures commonly occur where the soil profiles are thin in relation to the slope length, in particular where there are clearly–defined horizons having distinctive hydrological and geotechnical behaviour. Since this forms the major focus of the investigations into Holocene slope failure described above, where the progressive development of differentiated horizons leads to shallow translational failure, then the infinite planar stability analysis is appropriate. It has the added advantage of being relatively simple to parameterise.

However, the central issue emerging from research to date involves the role of unsaturated conditions and negative pore water pressures in the stability analysis. Negative pore water pressures impart additional resistance to the soil compared with the zero or positive pore water pressures traditionally emphasised in long–term slope stability studies. In most stability analyses, and in previous model applications to Holocene slope instability, negative pore water pressures are incorporated directly into the Mohr–Coulomb shear resistance equation, described as follows:

$$s = c' + (\sigma - \mu) \tan \phi' \tag{1}$$

where

s = shear resistance,

c' = effective cohesion,

σ = normal force,

μ = pore water pressure,

ϕ' = angle of internal friction with respect to effective stresses.

Where pore water pressures are negative (suctions), the effective normal force is enhanced, providing additional resistance to movement. It has been suggested that this is an inadequate representation of the role of negative pore water pressures since it ignores the role of pore air pressures (μ_a) which become particularly significant where soil suctions become high (Fredlund et al. 1978, Anderson et al. 1987). To refine the equations to define shear resistance involves consideration of three independent stress state variables $(\sigma - \mu_w)$, $(\sigma - \mu_a)$ and $(\mu_a - \mu_w)$, from which two equations of shear resistance can be derived (Fredlund et al. 1978). These equations are described as follows:

$$s = c' + (\sigma - \mu_w) \tan \phi' + (\mu_a - \mu_w) \tan \phi'' \tag{2}$$
$$s = c' + (\sigma - \mu_a) \tan \phi' + (\mu_a - \mu_w) \tan \phi^b \tag{3}$$

A three–dimensional plot enables calculation of the shear resistance for any suction (Fig. 8). It can be seen that as the suction increases, so the cohesion becomes greater. The rate of increase in cohesion as suction increases is described by the ϕ^b term, according to equation (3).

The two methods for suction inclusion were compared using an infinite planar slope stability analysis, assuming a ϕ^b value of 15° (Fredlund et al. 1978,

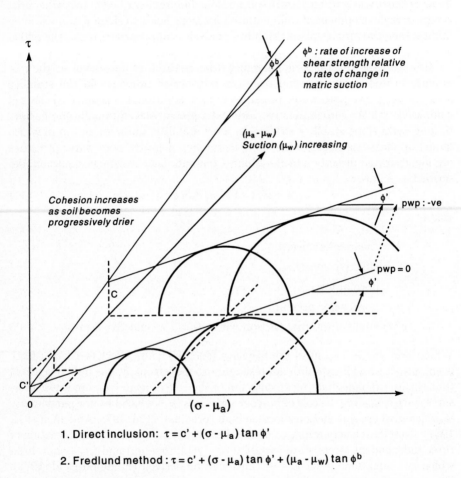

1. Direct inclusion: $\tau = c' + (\sigma - \mu_a) \tan \phi'$

2. Fredlund method : $\tau = c' + (\sigma - \mu_a) \tan \phi' + (\mu_a - \mu_w) \tan \phi^b$

ϕ^b ranges from $10° - 20°$

Fig. 8. Fredlund (1980) method for describing unsaturated zone shear resistance.

Fredlund 1980, 1987) and a slope angle of 30°. With direct inclusion (equation 1) large increases in the factor of safety accompany a change in suction from 0 to 20 kPa. Using the Fredlund method (equation 3) there are also increases in the factor of safety as suction increases, but these are not as marked as with direct inclusion of suction. The implication is that by failing to include the Fredlund method, slope failure will not be predicted in certain situations where it is actually likely to occur.

Generally, differences only become significant above suctions of around 5 kPa, although suctions of 10 kPa are reported in Anderson et al. (1987). However, variation is to be expected depending on the frictional properties of the material and the depth of the likely failure surface. It is for shallow depths and high suctions that the greatest differences are found. In the examples considered in this paper, "worst case" scenarios from resistance envelope calculations involve suctions of 5 kPa, and those found from laboratory monitoring involve 8 kPa. With failure occurring at about 80 cm below the ground surface, the difference between the methods is considerable. Accordingly, an investigation was carried out to consider angles of limiting stability associated with suctions of 5 and 10 kPa. Fig. 9 indicates that there can be up to a 6° over–estimation in angles of limiting stability resulting from failure to represent adequately unsaturated zone shear resistance. This is significant when compared with differences in angles of limiting stability reported for progressive pedogenesis (Brooks et al. 1993), which are of the order of 10°, as well as for changes associated with storm type (Brooks and Richards 1994), where angles of limiting stability vary by just a few degrees.

More recent applications have shown vegetation change to be less significant to slope instability than either climate or pedogenesis (Brooks and Collison 1996). However, as described above, routines for vegetation inclusion require refinement. Where transpiration extracts moisture and flux is upward during interstorm periods, the net effect is likely to be attainment of higher suctions. In such cases, as well as representation of the hydrological processes, the above analysis has shown that it is paramount to consider modification to the slope stability analysis. This is potentially of greater significance than simply addressing issues of scale or dimensionality in future applications of physically–based models to Holocene slope stability analysis.

Conclusions

Slope stability investigations for the Holocene have been greatly enhanced by the application of physically–based models in recent years. New light has been shed on the main temporal trends involved, permitting close consideration of the role of progressive soil development in the creation of conditions conducive to slope failure. While climatic change has not been discounted as an important influence on slope instability, the picture has been shown to be more complex than invoking a straightforward link between increased rainfall totals and shallow translational failure. Of central importance is the role of unsaturated zone

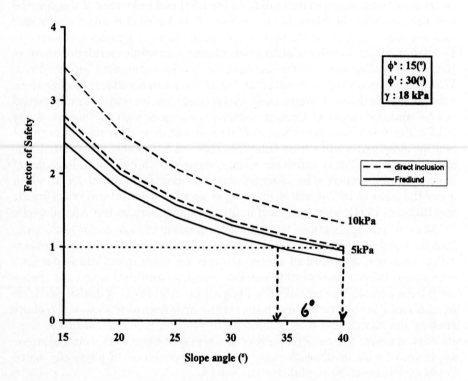

Fig. 9. Angles of limiting stability for low suctions using both direct inclusion of suction in the shear resistance equation and the Fredlund (1980) method.

hydrology in developing soil profiles, and the role that negative pore water pressures play in dictating limiting stable slope angles.

Focusing in some detail on process representation indicates that important factors in Holocene slope stability are inadequately represented by state–of–the–art models. In particular, representation of unsaturated zone shear resistance through direct inclusion of suctions in the Mohr–Coulomb shear resistance equation might lead to serious over–prediction of the limiting stable slope angles. An alternative approach suggested by Fredlund (1980) seems to provide direction for future model developments. Compounding the problem is the fact that existing vegetation routines probably fail to capture true "worst case" conditions. Existing routines do not allow for transpiration uptake of water by the vegetation and they represent root reinforcement simplistically. As a consequence, application of models to consider the role of vegetation in slope failure has shown it to be of limited significance compared with climate or with progressive pedogenesis. If we take into account the fact that higher suctions are likely to result from better representation of hydrological and geotechnical processes associated with the presence of vegetation, and that this then makes the failure to utilise the Fredlund method of stability analysis more serious, it is clear that the effect of vegetation change on slope failure in the Holocene is only partially understood. New research should address this issue as a priority.

There has been an increasing trend in recent years towards model development to larger scales and increasingly detailed geometric representation. In terms of research into Holocene slope stability, there is undoubtedly scope for future application of such models where saturated conditions develop and lateral throughflow becomes important. However, without adequate process representation for the unsaturated zone, slope stability involving recently developed soil profiles with a developing vegetation cover cannot be considered in detail, and it is felt that this needs to form a major future research area. Once detailed process mechanisms are understood, the most significant factors to slope stability in the long–term can be identified. This will then provide a platform for model development where long–run simulations can be prioritised. In such models, detailed focus on process mechanisms is precluded, and reliance needs to be placed on small scale models to set the agenda for less detailed process representation in long–term models. This combination of detailed process–based models and lumped long–term models can provide a powerful means of assessing long–term changes in slope stability.

References

Anderson, M.G., K.S. Richards, and P.E. Kneale (1980): The role of stability analysis in the interpretation of the evolution of threshold slopes. Trans. Inst. Brit. Geog. 5: 101–112.

Anderson, M.G., M.J. Kemp, and J.M. Shen (1987): On the use of resistance envelopes to identify the controls on slope stability in the tropics. Earth Surface Processes and Landforms 12: 637–648.

Anderson, M.G., K.S. Richards, and P.E. Kneale (1980): The role of stability analysis in the interpretation of the evolution of threshold slopes. Trans. Inst. Brit. Geog. 5: 101–112.

Anderson, M.G., M.J. Kemp, and D.M. Lloyd (1988): Refinement of hydrological factors for the design of cut slopes in the tropics. International Society for Soil Mechanics and Foundation Engineering Proceedings: 233–240.

Anderson, M.G., A.J.C. Collison, J. Hartshorne, D.M. Lloyd, and A. Park (1986): Developments in slope hydrology–stability modelling for tropical slopes. In: Anderson, M.G., and S.M. Brooks (eds.) Advances in Hillslope Processes, Wiley, Chichester: 799–822.

Beven, K.J. (1989): Changing ideas in hydrology – the case of physically based models. Jnl. Hydrol. 105: 157–172.

Beven, K.J., A.M. and Binley (1992): The future of distributed models: model calibration and uncertainty prediction. In: Beven, K., and I.D. Moore (eds.) Terrain analysis and distributed modelling in hydrology, Wiley, Chichester: 227–246.

Binley, A.M., K.J. Beven, A. Calver, and L.G. Watts (1991): Changing responses in hydrology: assessing the uncertainty in model predictions. Water Resources Research 27: 1253–1261.

Bockheim, J.G. (1980): Solution and use of chronofunctions in studying soil development. Geoderma 24: 71–85.

Bishop, A.W. (1955): The use of the slip circle in the stability analysis of slopes. Geotechnique 5: 7–17.

Brazier, V., and C.K. Ballantyne (1989): Late Holocene debris cone evolution in Glen Feshie, western Cairngorm Mountains, Scotland. Trans. Roy. Soc. Edinburgh, Earth Sciences 80: 17–24.

Brooks, S.M., and K.S. Richards (1994): The significance of rainstorm variations to shallow translational hillslope failure. Earth Surface Processes and Landforms 18: 85–94.

Brooks, S.M., and M.G. Anderson (1995): The determination opf suction–controlled slope stability in humid temperate environments. Geografiska Annaler 77A: 11–22.

Brooks, S.M., and A.J.C. Collison, (1996): The significance of soil profile differentiation to hydrological response and slope instability: a modelling approach. In: Anderson, M.G., and S.M. Brooks (eds.) Advances in Hillslope Processes, Wiley, Chichester: 471–486.

Brooks, S.M., K.S. Richards, and M.G. Anderson (1993): Shallow failure mechanisms during the Holocene: utilisation of a coupled soil hydrology–slope stability model. In: Thomas, D., and R. Allison (eds) Landscape Sensitivity, Wiley, Chichester: 149–174.

Brooks, S.M., M.G. Anderson, and A.J.C. Collison (1995a): Modelling the role of climate, vegetation and pedogenesis in shallow translational hillslope failure. Earth Surface Processes and Landforms 20: 231–242.

Brooks, S.M., M.G. Anderson, and K. Crabtree (1995b): The significance of fragipans to early–Holocene slope failure: application of physicaly–based modelling. The Holocene 5: 293–303.

Caine, N. (1980): The rainfall intensity–duration control of shallow landslides and debris flows. Geografiska Annaler 62A: 23–27.

Campbell, G.S. (1974): A simple method for determining unsaturated conductivity from moisture retention data. Soil Science 117: 311–314.

Carson, M.A. (1975): Threshold and characteristic angles of straight slopes. Proc. 4th Int. Symp. on Geomorph.: 19–34.

Carson, M.A., and D.J. Petley (1970): The existence of threshold slopes in the denudation of the landscape. Trans. IBG 49: 71–95.

Collison, A.J.C. (1995): Assessing the influence of vegetation on slope stability in the tropics. PhD thesis, University of Bristol.

DeRoo, A.P.J. (1996): Validation problems of hydrologic and soil erosion catchment models: examples from a Dutch erosion project. In: Anderson, M.G., S.M. and Brooks (eds.) Advances in Hillslope Processes, Wiley, Chichester: 669–684.

Fawcett, K.R., M.G. Anderson, P.D. Bates, J.-P. Jordan, and J.C. Bathurst (1995): The importance of internal validation in the assessment of physically–based distributed models. Trans. Inst. Brit. Geog. NS 20: 248–265.

Feddes, R.A., P. Kabat, P.J.T. van Bakel, J.J.B. Bronswijk, and J. Halbertsma (1988): Modelling soil water dynamics in the unsaturated zone – state of the art. Journal of Hydrology 100: 69–111.

Fredlund, D.G., N.R. Morgenstern, and R.A. Widger (1978): The shear strength of unsaturated soils. Canadian Geotechnical Jnl. 15: 313–321:

Fredlund, D.G. (1980): The shear strength of unsaturated soil and its relation to slope stability problems in Hong Kong. Hong Kong Engineer 8: 57–59.

Fredlund, D.G. (1987): Slope stability analysis incorporating the effect of soil suction. In: Anderson, M.G., and K.S. Richards (eds.) Slope Stability: geotechnical engineering and geomorphology. Wiley, Chichester: 113–144.

Freeze, R.A. (1987): Modelling interrelationships between climate, hydrology and hydrogeology and the development of slopes. In: Anderson, M.G., and K.S. Richards (eds.) Slope Stability: geotechnical engineering and geomorphology. Wiley, Chichester: 381–403.

Grayson, R.B., I.D. Moore, and T.A. McMahon, T.A. (1992): Physically based hydrological modeling 2. Is the concept realistic? Water Resources Research 26: 2659–2666.

Greenway, D.R. (1987): Vegetation and slope stability. In: Anderson, M.G., and K.S. Richards (eds.) Slope stability: geotechnical engineering and geomorphology. Wiley, Chichester: 187–230.

Grove, J.M. (1972): The incidence of landslides, avalanches and floods in western Norway during the Little Ice Age. Arctic and Alpine Research 4: 131–138.

Innes, J.L. (1983): Lichenometric dating of debris flow deposits in the Scottish Highlands. Earth Surface Processes and Landforms 8: 579–588.

Janbu, N. (1954): Stability analysis of slopes with dimensionless parameters. Harvard University Soil Mechanics, Series No 46.

Konikow, L.F., and J.D. Bredehoeft (1992): Ground–water models cannot be validated. Advances in Water Resources 15: 75–83.

Millington, R.J., and J.P. Quirk (1959): Permeability of porous media. Nature 183: 387–388.

Nash, D.F.T. (1987): A comparative review of limit equilibrium methods of slope stability analysis. In: Anderson, M.G., and K.S. Richards (eds.) Slope stability: Geotechnical engineering and geomorphology. Wiley, Chichester: 11–76.

Running, S.W., and J.C. O'Loughlin (1988): A general model of forest ecosystem processes for regional applications. I. Hydrologic balance, canopy gas exchange and primary production processes. Ecological modelling 42: 125–154.

Sidle, R.C. (1991): A conceptual model of changes in root cohesion in response to vegetation management. Jnl. Env. Quality 20: 43–52.

Sidle, R.C. (1992): A theoretical model of the effects of timber harvesting on slope stability. Water Resources Research 28: 1897–1910.

Sissons, J.B. (1979): The Loch Lomond Stadial in the British Isles. Nature 278: 199–203.

Starkel, L. (1966): Postglacial climate and the moulding of European relief. Proc. Int. Symp. World Climate 8000–0BC, Royal Met. Soc. London: 15–32.

Van Genuchten, M.Th. (1980): A closed–form equation for predicting the hydraulic conductivity of unsaturated soils. Soil Science Society of America Journal 44: 892–898.

Waldron, L.J., S. Dakessian, and J.A. Nemson (1983): Shear resistance enhancement of 1.22–meter diameter soil cross sections by pine and alfalfa roots. Soil Science soc. Am. Jnl. 47: 9–14.

Wu, W., R.C. and Sidle (1995): A distributed slope stability model for steep forested basins. Water Resources Research 31: 2097–2110.

Young, J.A.T. (1971): The terraces of Glen Feshie, Inverness–shire. Trans. Roy. Soc. Edinburgh 69: 500–512.

Self–Organized Criticality in Landsliding Processes

St. Hergarten and H. J. Neugebauer

Geodynamics – Physics of the Lithosphere, University of Bonn, Germany

Abstract. Using a physically based model we show that landsliding may be seen as a self–organized criticality (SOC) process if the long term driving forces (fluvial or tectonic) are regarded. The model is based on partial differential equations and combines aspects of slope stability and mass movement. With its help, the scale invariant frequency magnitude relations observed in many regions can be understood and probably be predicted; temporal behaviour and landslide geometry can be investigated statistically. Moreover, we discuss the effects of SOC on landforms being out of equilibrium due to changing terrain resistance or climatic conditions.

1 Introduction

In the last years, a lot of research has been done on scale invariance and scale dependence of the land surface (e. g. Evans 1998, Kirkby 1998). More than 50 years ago, Horton (1945) found scale invariance in river networks. Later, scale invariance of landslides was discovered in field (Fuyii 1969, Sugai et al. 1994, Yokoi et al. 1995, Goltz 1996, Hovius et al. 1997) as well as in laboratory experiments (Somfai et al. 1994).

The main result of these studies is that the frequency of landslide occurence as a function of their magnitude (i. e., the affected area) can often be described by a power law. Among all statistical distributions, the power law is the only scale invariant distribution; it links the probability of large and small objects without emphasizing any object size. Power law distributions are also called fractal distributions. It should be noted that for physical as well as for mathematical reasons (normalization) a power law distribution can only be valid over a limited range; outside this range scale invariance breaks down, e. g. Hovius et al. (1997) established a power law distribution for landslides with areas between $5000\,\mathrm{m}^2$ and $1\,\mathrm{km}^2$. Recent discussion about scale invariance or scale dependence of the land surface mostly concerns the question about the range where scale invariance applies and where it breaks down.

Even if distributions of event sizes themselves are important and interesting, our view of nature is mostly based on our understanding of the processes that generate the patterns we observe in field. From this point of view, establishing the concept of self–organized criticality (SOC) was a major step to understand fractal distributions. The concept of SOC was introduced first by Bak et al.

(1987,1988); since then it has been applied to phenomena in physics, biology, economics, and geosciences (Bak 1996).

A system exhibits SOC behaviour if it

(a) tends to move into a quasi stationary state, where
(b) the distribution of event sizes (e. g., the affected area) is scale invariant, and where
(c) the temporal behaviour is a $\frac{1}{f}$ (pink, flicker) noise.

In this case, the quasi stationary state is called critical state.

Concerning landform evolution, the occurence of fractal structures suggests SOC behaviour, but establishing the SOC concept in landform evolution theory is still difficult. Because time scales are long, field data often do not reveal sufficient temporal data, so that they only yield the spatial distribution (b). The discussion initiated by Sapozhnikov and Foufoula–Georgiou (1996) addresses the same aspect in current landform models; e. g., the erosion models of Takayasu and Inaoka (1992) and Rinaldo et al. (1993) are able to reproduce the fractal characteristics of river networks, but finally lead to stationary flow patterns. The latter effect is clearly inconstistent with the SOC concept that predicts a permanent fluctuations around a long term equilibrium in agreement with our experience of nature. Lately, strong hints for the occurence of SOC in the evolution of braided rivers were found by Sapozhnikov and Foufoula–Georgiou (1997) from laboratory experiments.

In the following, we concentrate on a landslide model developed recently by Hergarten and Neugebauer (1998). Apparently, it was the first landslide model that exhibits proper SOC behaviour. In the next session, we recapitulate and discuss the basics of the model and the derivation of the model equations. The third section illustrates the concept of SOC und its implications on landslide dominated landform evolution using the model. In the fourth section the geometric properties of the landslides generated by the model are discussed. Finally, in the fifth section we briefly discuss the effect of SOC on landforms being out of equilibrium due to changing terrain resistance or climatic conditions.

2   The Model

In order to investigate a phenomenon with respect to SOC, a model must be suitable for simulations over long periods where thousands of events occur. In the special case of landslides, aspects of slope stability (Brunsden and Prior 1984, Craig 1992) and granular flow (debris flow) (Iverson 1997, Jan and Shen 1997) have to be combined. Thus, the approach should be quite simple compared with models dealing with one aspect in detail.

It is well known that slope stability does not only depend on slope gradient; the meachanical state of the soil has to be considered, too. Hence we need at least one additional space and time dependent variable beside the surface height $H(x_1, x_2, t)$. We characterize the mechanical state of the soil in a simple way assuming that the soil consists of two layers: a lower one that is tight and an upper one that is able to flow or creep if appropriate driving forces (slope gradient) apply (Fig. 1). A comparable 'two state' description was developed by Bouchaud et al. (1995) for modeling sandpile dynamics; they introduced a density of rolling grains as a second variable beside the surface height. Hergarten and Neugebauer (1996) used the density of loose particles lying on top of the land surface in an erosion model.

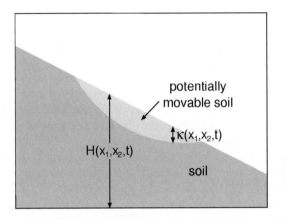

Fig. 1. Coordinates and variables.

Let $\kappa(x_1, x_2, t)$ be the thickness of the upper, movable layer. Instability may be induced by a large slope gradient or by a large amount of movable material (Hutchinson and Bhandari 1971). The simplest criterion regarding both, is that instability occurs if the product $\kappa|\nabla H|$ exceeds a given threshold β. In this case, we assume a linear flow behaviour driven by the excess of $\kappa|\nabla H|$ vs. β directed downslope, although it is known that the real behaviour of debris flow may be more complicated (Iverson 1997, Jan and Shen 1997). Thus, the flow density is:

$$j = \begin{cases} -\alpha\left(\kappa|\nabla H| - \beta\right)\frac{\nabla H}{|\nabla H|} & \text{if} \quad \kappa|\nabla H| > \beta \\ 0 & \text{else} \end{cases} \tag{1}$$

The parameter α can be interpreted as the flow velocity of a thick layer ($\kappa \gg \beta$) on a slope of 100 % inclination. At this place, it should be mentioned that our approach does not include any effects of inertia, so that it describes slow creep rather than rapid flow.

Inserting expression (1) into the equation of continuity (mass conservation) yields a partial differential equation for the surface height:

$$\frac{\partial H}{\partial t} = \text{div}\begin{cases} \alpha\left(\kappa|\nabla H| - \beta\right)\frac{\nabla H}{|\nabla H|} & \text{if} \quad \kappa|\nabla H| > \beta \\ 0 & \text{else} \end{cases} \tag{2}$$

In soil mechanics, slope stability is often based on the concept of the factor–of–safety which balances stabilizing and driving forces. Our approach does not

contain a factor–of–safety explicitly, but can in principle be related to it: if
the slope is not too steep, the factor–of–safety can in general be interpreted as
the factor by which the slope could be steepened until instability occurs. As
mentionend above, in our model instability occurs if the ratio

$$S = \frac{\beta}{\kappa\,|\nabla H|} \tag{3}$$

is smaller than one. Thus, we can interpret S as a local factor of stability.

The following, second equation shall describe how the soil properties change,
i.e., how the boundary $H - \kappa$ between tight and movable soil is shifted. First,
the effects of infiltration of water and subsurface flow on slope stability have to
be regarded; but they are quite complicated and require at least one additional
(hydrological) equation. Instead, we summarize them using a random impact
$r(x_1, x_2, t)$ that increases the amount of movable material. Describing the growth
of layer thickness, r has the dimension length per time. Due to the episodic
behaviour of rainfall events and the inhomogeneous distribution of hydraulic
conductivity and fissures, r varies temporally and spatially. The opposite process
– stabilization by drying – is assumed to be linear with a characteristic time τ.
Hence the amount of movable material decays exponentially if there is no impact
by infiltration of water. Moreover, it is well known that moving material may
damage tight soil, so that the amount of sliding mass may increase downslope. In
order to regard this, we assume that a part of the energy dissipated by the flow
makes material movable. In order to quantify this effect, we consider an arbitrary
section Ω of the x_1–x_2 plane; the potential energy of the material above Ω is

$$E = \int_\Omega \frac{1}{2}\,H^2 \tag{4}$$

where the specific weight of the soil has been normalized to one. Using the
equation of continuity and the theorem of Gauss, we obtain:

$$\frac{\partial E}{\partial t} = -\int_\Omega j \cdot \nabla H - \int_{\partial\Omega} jH \tag{5}$$

Thus, $-j \cdot \nabla H$ is the energy dissipation and jH is the flux of potential energy.

Summarizing the three effects described above und inserting equation (1),
we obtain the second differential equation:

$$\frac{\partial}{\partial t}(H - \kappa) = -r + \frac{\kappa}{\tau} \underbrace{- \gamma\alpha\,(\kappa|\nabla H| - \beta)\,|\nabla H|}_{\text{if } \kappa|\nabla H| > \beta} \tag{6}$$

The parameter γ with the dimension of an inverse length quantifies the effect of
the energy dissipation.

The coupled system with the variables H and κ is solved numerically using a
finite volume method with an implicit time discretization and a fixed point iteration scheme. In the first half step of each iteration cycle, a new approximation

to κ is calculated by solving the difference of the equations (2) and (6) using a simple method of characteristic curves (Enquist–Osher scheme) (Kröner 1997). In the second half step a new approximation to H is calculated by inserting the surface H from the last cycle into the right hand side of equation (2). For the reason that the convergence rate of the iteration decreases significantly during large landslide events, the algorithm requires an adaptive time step variation.

3   A Numerical Example

Landslides are dissipative phenomena; after the material has come to rest, its potential energy has decreased compared to the initial state. In general, dissipative systems reach a stationary state or tend towards a stationary state in absence of external driving forces. In contrast, the SOC concept requires that the surface permanently undergoes alterations, so that SOC behaviour can only be obeserved if the long term driving forces are regarded, too. Concerning landslides, the driving forces are

- the downcutting of rivers that steepens the slope feet on the catchment scale and
- tectonic uplift processes on regional to continental scales.

In the example shown here, we consider the simplest case of a river downcutting at the foot of a single slope. This example can be seen as a basic case of landslide dominated landform evolution on the catchment scale. The downcutting itself is not modeled, but parametrized by a time dependent Dirichlet boundary condition with a uniformly decreasing boundary value for the surface height. The other boundary conditions are illustrated in Fig. 2.

The parameter values are: $\alpha = 20\,\mathrm{m/yr}$, $\beta = 1\,\mathrm{m}$, $\gamma = 0.1\,\mathrm{m}^{-1}$, and $\tau = 1\,\mathrm{yr}$; the downcutting rate is $1\,\mathrm{cm/yr}$. The random impact $r(x_1, x_2, t)$ is assumed to be constant over periods of 0.1 yrs. We consider this time span as a reasonable duration of wet periods which are able to destabilize slopes. For each period of 0.1 yrs, a random value r_0 is determined from an exponentially decreasing distribution with an average of $1\,\mathrm{m/yr}$. In a second step, $r(x_1, x_2, t)$ is calculated for every cell of our finite volume grid ($16 \times 16\,\mathrm{m}^2$) by applying a small, uniformly distributed random variation of maximal $\pm10\,\%$ to r_0. The average value can be interpreted in terms of recurrence of wet periods; solving equation (6) in the stable case (third term vanishes) shows that a wet period that destabilizes a slope of $75\,\%$ inclination occurs every 10 years in the mean. A slope of $50\,\%$ inclination is destabilized once in 300 years.

Fig. 3 illustrates the first visible effects of river downcutting. The upper part shows that the slope foot has been steepened, a minor soil creep has occured. About 50 years later, considerable landslides have roughened the slope foot significantly. Finally, the lower part illustrates that these landslides are not individual ones but built up a large one that affects the whole slope and leads to an enormous soil loss.

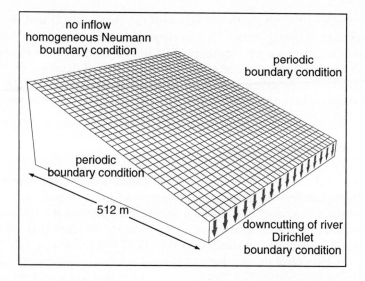

Fig. 2. Initial situation with boundary conditions.

Fig. 4 quantifies this behaviour; the average loss of height is about 40 m, corresponding to a volume of 10^7 m^3. It also shows that such a large event does not occur any more during the rest of the simulation. After about 10000 years, a quasi stationary equilibrium between river downcutting and landslide caused soil loss. Fluctuations of different sizes occur.

Fig. 5 shows some snapshots taken from the quasi stationary state; landslide scars of different sizes are visible. The largest events (Fig. 5) affect nearly the whole slope.

The cumulative frequency magnitude relation of the events in the quasi stationary state is plotted in Fig. 7. It quantifies the number of landslides larger than a given size (affected area) per time and total area considered as a function of the event size. Over three orders of magnitude in area, the distribution follows a power law as required for SOC. The straight line shows the best fit power law, fitted to the non–cumulative distribution with logarithmic bin widths in order to avoid statistical bias. The few large events above 50, 000 m^2 do not affect the exponent significantly. In coincidence with the landslide mapping studies of Fuyii (1969) and Hovius et al. (1997), who obtained exponents of 0.96 resp. 1.16, our numerical example yields an exponent of 0.99. The fact that our absolute sliding frequencies are by a factor 5–10 higher than those of Hovius et al. (1997) should not be overinterpreted before detailed parameter studies have been performed.

In order to proof that the quasi stationary state is critical according to the SOC concept, the temporal scaling behaviour has to be analyzed, too. In the first paper describing this model (Hergarten and Neugebauer 1998), it has been stated that the sediment yield per time into the river is a pink noise. Instead, in this paper we analyze the scaling properties of landslide duration directly (Fig. 8). Over one and a half orders of magnitude, the distribution is scale

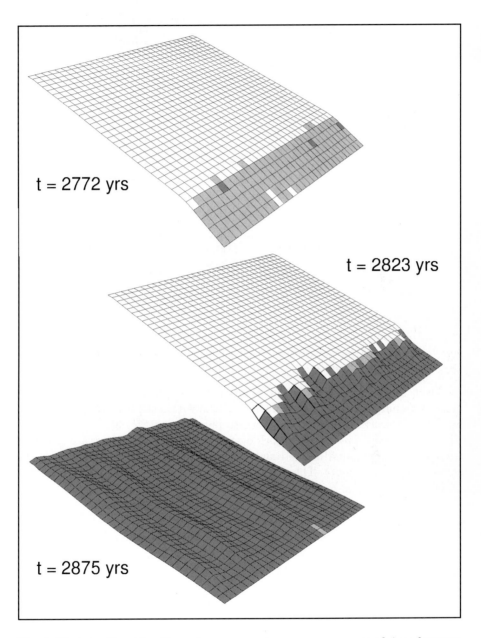

Fig. 3. First significant landform changes occuring as a consequence of river downcutting. Regions that became unstable during the last 50 years are marked with light gray; regions where a significant loss of height occured (larger than the downcutting of the river) are marked with dark gray.

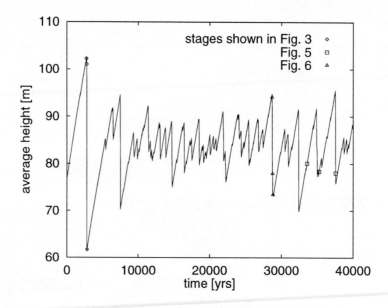

Fig. 4. Average surface height (relative to the slope foot) vs. time.

invariant, even if the power law is not determined as well as the power law obtained for the affected area.

Thus, we have shown that our model exhibits proper SOC behaviour; the quasi stationary state can be considered critical within the SOC framework. But how does it work? Why did the 'catastrophic' event (Fig. 3) occur so early? The SOC concept predicts that such events should be rare. The reason for this behaviour can be found in the roughness of the surface. In the beginning, the surface was quite smooth. The random impact that parametrizes the infiltration of water hardly varies spatially, so that the variations of stability on the smooth surface are small, too. If any instability occurs, it will affect the neighbourhood easily, so that the event grows infinitely. Afterwards, the surface has been roughened; this causes significant spatial variations in stability. Fig. 9 shows the probability distribution of the local factor of stability (equation 3), taken over the whole slope in the critical state.

In the mean, only 1 % of the slope area is unstable; 87.5 % of the slope have a local factor of stability above 1.5 in our example. This distribution shows that the critical state is not what we would imagine intuitively – a state where any small disturbation has a great effect, like the one shown in Fig. 3. In contrast, the critical state is a state with a complex spatial structure; it is organized in a way that a small disturbation may cause a large landslide, but usually it does not. The roughness allows the occurence of many modest size events; the probabilities of events of different sizes are linked by a scale invariant frequency magnitude relation.

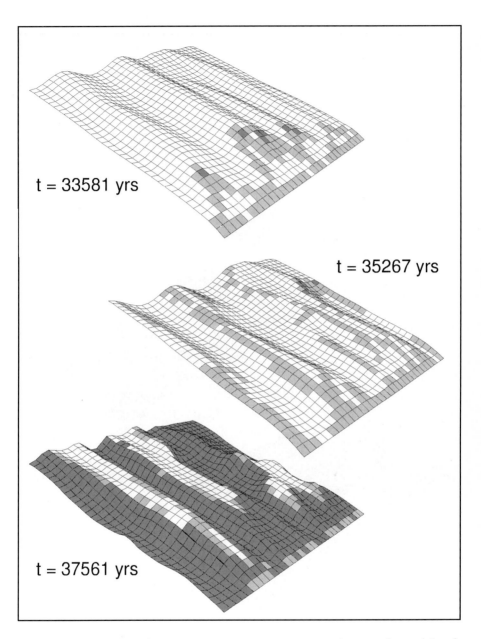

Fig. 5. Snapshots taken from the quasi stationary state: a quite smooth one (above), a medium rough one (middle), and a quite rough one (below). Regions that became unstable during the last 50 years are marked with light gray; regions where a significant loss of height occured (larger than the downcutting of the river) are marked with dark gray.

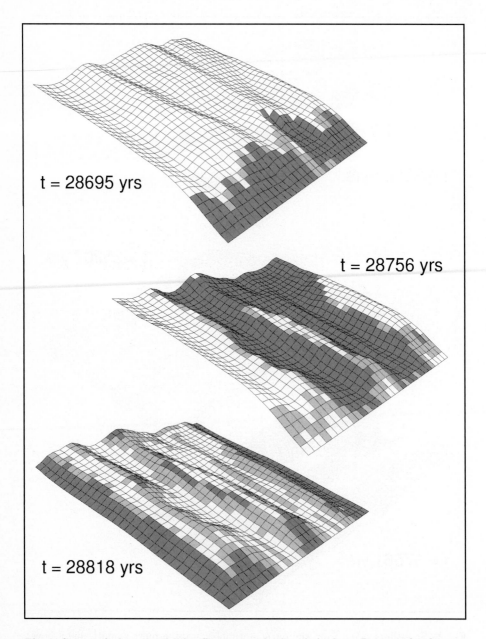

Fig. 6. Stages of a large landslide affecting nearly the whole slope. Regions that became unstable during the last 50 years are marked with light gray; regions where a significant loss of height occured (larger than the downcutting of the river) are marked with dark gray.

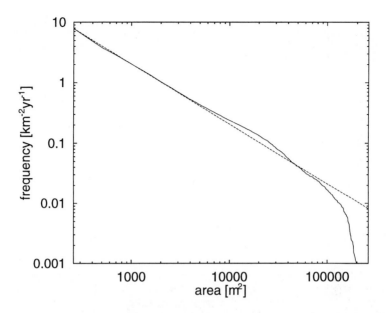

Fig. 7. Cumulative frequency magnitude relation, i. e., number of landslides larger than a given area per time and considered area. The power law with an exponent of 0.99 was fitted over the whole range of events.

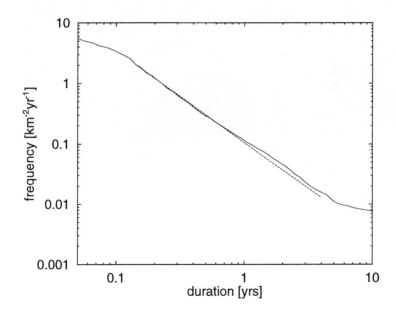

Fig. 8. Number of events exceeding a given duration per time and considered area. The power law with an exponent of 1.52 was fitted between 0.14 and 4 years.

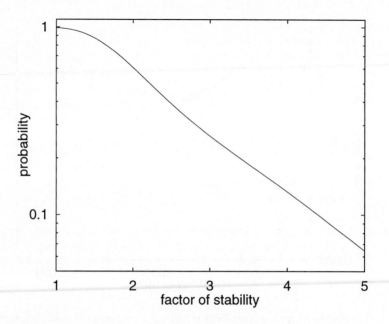

Fig. 9. Probability distribution of the local factor of stability.

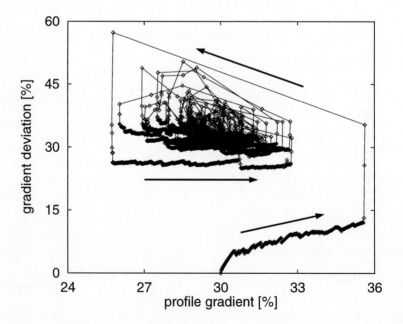

Fig. 10. Simplified phase space diagram. The arrows indicate the temporal evolution; the nodes on the line mark 10 year intervals.

The most striking property of SOC systems is their tendency to organize themselves into the critical state independently of the initial condition; they are able to generate the complex spatial structure they need by themselves. Fig. 10 illustrates how our landslide model organizes the surface roughness. The gradient of a straight profile fitted to the slope is plotted versus the quadratically averaged (over the whole slope) deviation of the surface gradient from the profile. Fig. 10 can be interpreted as a simplified phase space diagram where the phase space consists of steepness and roughness. For the reason that the system tends to the critical state indepently of the initial conditions, the critical state is an attractor in phase space. The dots on the curve indicate that there are long periods where the surface is not altered significantly, except for some steepening due to the river downcutting. Compared to these periods, large events are quite short. In a first step, the roughness increases as expected; then an additional decrease of profile gradient occurs due to the soil loss into the river. In a third phase, smaller landslides occuring at the edges of the main slide smooth out the surface again. The outer, large loop corresponds to the large event occuring quite early (Fig. 3). It can clearly be seen, that the system organizes itself into a state where the profile gradient varies between 25 % and 33 %, and the roughness (gradient deviation) varies between 30 % and 50 % in this example; but these values are expected to depend on mechanical properties, climatic conditions, and boundary conditions.

4 Landslide Geometry

In the previous section, we considered the affected area as a measure of landslide magnitude. In geomorphology, lots of other geometric properties are used for characterizing landslides, e. g., length, width, thickness, and volume. Except for the volume, that is usually considered as the total volume of soil moved by the landslide, their definition is not unique. We define

- the length as the maximum extension of the affected area in direction of the slope,
- the width as the ratio of affected area and length, i. e., as an average width, and
- the depth as the ratio of volume and affected area, i. e., as an average depth.

If we analyze these geometric properties statistically, a problem arises from the frequency magnitude relation (Fig. 7) which decreases approximately with the inverse of the area, so that the results are governed by the small events. In order to eliminate this effect, we use a statistical measure which weights landslides according to their area. Using this measure, all the small events together are weighted equally to all large events together.

Table 1 lists the mean values and standard deviations of the landslides' geometric properties obtained from the model. In agreement with the results of Hovius et al. (1997), the average values of length and width suggest that landslide width is about half of landslide length in the mean.

Table 1. Mean values and standard deviations of the landslides' geometric properties.

	length [m]	width [m]	thickness [m]	area [m^2]	volume [m^3]	duration [yrs]
mean value	243	133	0.79	41900	112000	3.1
standard deviation	201	117	3.1	55200	587000	9.7

Table 2 shows the correlation coefficients between the geometric properties obtained from the model. It shows that most of the properties are poorly correlated with coefficients below 0.5. The weak correlation of length and width indicates that the ration of length and width obtained from the mean values should not be overinterpreted. The data must be expected to be scattered widely around this ratio. Fig. 11 confirms this suspicion: length and width seem to be distributed nearly independently of each other; there is no preferred region with a constant aspect ratio which would occur as a dark straight line between the axes. Hence there is a great variety of landslide shapes in the model; long narrow slides occur as well as short wide landslides. The fractal distribution of landslide areas does not imply that all slides look are nearly similar in shape.

Table 2. Correlation coefficients between the landslides' geometric properties.

	length	width	thickness	area	volume	duration
length		0.40	0.29	0.77	0.25	0.35
width	0.40		0.34	0.75	0.35	0.37
thickness	0.29	0.34		0.46	0.94	0.92
area	0.77	0.75	0.46		0.47	0.51
volume	0.25	0.35	0.94	0.47		0.82
duration	0.35	0.37	0.92	0.51	0.82	

The better correlation between length and area, resp. width and area, should not be overinterpreted, too. They can be considered as artificial effects because area is the product of length and width.

If the effect of landsliding on landform evolution shall be quantified, the landslide volume is the most significant variable. Its correlation with length, width, and area is weak; it seems that landslide volume is governed by landslide depth more than by length, width or area. Thus, estimation of landslide volume from these one or two dimensional properties should be treated very carefully. landslide depth itself seems to be correlated best with landslide duration.

In the model, the frequency of landslide occurence as a function of landslide volume can be computed directly (Fig. 12). Despite the poor correlation with landslide area, the distribution is a power law over some orders of magnitude, too. The exponent of the best fit power law is 0.54. The relation suggested by Ohmori (1992) and Hovius et al. (1997) claims that the exponent should be two

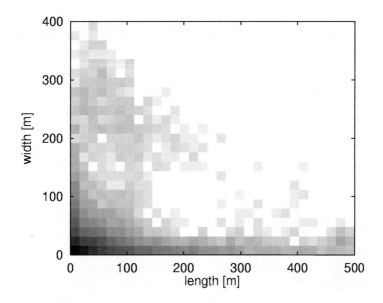

Fig. 11. Number of events as a function of their length and width. Dark areas indicate large number of events.

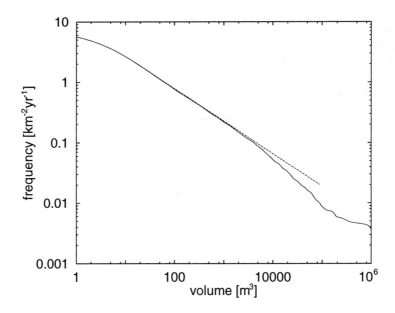

Fig. 12. Number of events exceeding a given volume per time and considered area. The power law with an exponent of 0.54 was fitted between 15 and 90,000 m³.

thirds of the exponent of the area distribution (Fig. 7), i. e. 0.66. However, our exponent is smaller than the one of Hovius et al. (1997), so that our simulated data even strengthen their result that landform evolution is governed by the few large events rather than by the large number of small events.

5   Landforms out of Equilibrium

The SOC concept basically depends on the long term equilibrium between driving forces and dissipative processes. For a basic understanding of the processes, this assumption is appropriate, but in nature the conditions cannot be expected to be constant over 10,000's of years. Climatic conditions change as well as the terrain resistance may change, e. g., due to anthropogenic influences like deforestation (Crozier and Preston 1998).

As an example, we assume that the threshold β is reduced by 20 % from 1 m to 0.8 m. For simplicity, this decrease of terrain resistance shall take place immediately at an arbitrary stage picked out of the critical state. Fig. 13 shows that the system leaves its previously preferred region of phase space, the critical state with reference to $\beta = 1$ m, and tends towards a new critical state. The critical state corresponding to the new parameter value $\beta = 0.8$ m is distinguished from the original one by a lower roughness and by slightly lower profile gradients. Apart from this, both critical states are similar; the exponent of the frequency magnitude relation (Fig. 14) decreases slightly, so that the probabilty of large landslides increases with decreasing threshold.

But what happens during the transition between the two states? In order to investigate this we performed 200 simulations of 100 years, beginning at stages randomly picked out of the original critical state. The resulting frequency magnitude relation is plotted as the third curve in Fig. 14; it is significantly higher than the frequency magnitude relations of the original critical state and of the new critical state the system tends to. Especially, the tail of the curve indicates that spatial scaling breaks down at large events; events which nearly affect the whole slope seem to be preferred. As an example, within the first 100 years after reducing the threshold the probability of a landslide larger than 150,000 m^2 increases by a factor 10; later it decreases again to a value less than twice as high as the orginal one.

Thus, the behaviour within the transition phase may be far outside the range suggested by the corresponding equilibrium states, so that the properties of these states cannot be used for estimating the behaviour during the transition phase. From this insight, questions concerning the duration of transition phases, dependent on the changing conditions, immediately arise; in many cases, process models will be an important key to their answers.

6   Conclusions

With the help of a physically based landform evolution model we have shown that landsliding may be seen as a SOC process if the long term driving forces are con-

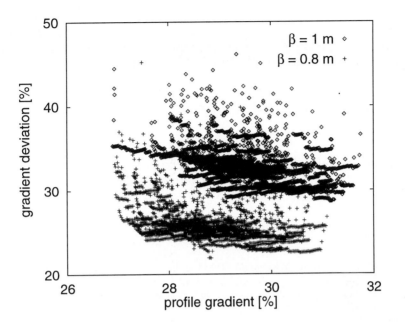

Fig. 13. Simplified phase space diagram. The arrows indicate the temporal evolution; the dots on the line mark 10 year intervals.

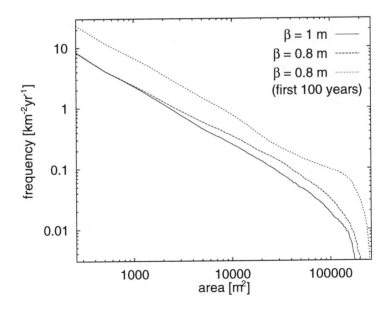

Fig. 14. Cumulative frequency magnitude relation, i. e., number of landslides larger than a given area per time and considered area. The power law with an exponent of 0.99 was fitted over the whole range of events.

sidered, too. Together with our knowledge of SOC processes, the model helps us to understand the often observed scale invariant frequency magnitude relations. Statistical analysis of the landslides' geometric properties reveals only weak correlation betweens length, width, and depth, corresponding to the great variety of landslide shapes observed in nature. Finally, it has been shown that changing conditions lead to transition states; their properties (e. g., their frequency magnitude relation) may deviate significantly from those of the corresponding equilibrium states.

Acknowledgment. The authors would like to thank the German Research Foundation (DFG) for financial support within the Collaborative Research Center (SFB) 350.

References

Bak, P., C. Tang, and K. Wiesenfeld (1987): Self–organized criticality. An explanation of 1/f noise. Phys. Rev. Lett., 59: 381–384.

Bak, P., C. Tang, and K. Wiesenfeld (1988): Self–organized criticality. Phys. Rev. A, 38: 364–374.

Bak, P. (1996): How nature works: the science of self–organized criticality. Copernicus, Springer, New York.

Bouchaud, J.-P., M.E. Cates, J. Ravi Prakash, and S.F. Edwards (1995): Hysteresis and metastability in a continuum sandpile model. Phys. Rev. Lett., 74: 1982–1985.

Brunsden, D., and B.D. Prior (1984): Slope instability. Wiley, Chichester.

Craig, R.F. (1992): Soil mechanics. 5th ed., Chapman & Hall, London.

Crozier, M.J., and N.J. Preston (1998): Modelling Changes in Terrain Resistance as a Component of Landform Evolution in Unstable Hill Country. In this volume.

Evans, I.S. (1998): Relations between land surface properties: altitude, slope and curvature. In this volume.

Fuyii, Y. (1969): Frequency distribution of the magnitude of landslides caused by heavy rainfall. Seism. Soc. Japan Journal, 22, 244–247.

Goltz, C. (1996): Multifractal and entropic properties of landslides in Japan. Geol. Rundschau, 85: 71–84.

Hergarten, S., and H.J. Neugebauer (1996): A physical statistical approach to erosion. Geol. Rundschau, 85:65–70.

Hergarten, S., and H.J. Neugebauer (1998): Self–organized criticality in a landslide model. Geophys. Res. Lett., 25(6): 801–804.

Horton, R.E. (1945): Erosional development of streams and their drainage basins; hydrophysical approach to quantitative morphology. Bulletin Geol. Soc. of America, 56:275–370.

Hovius, C., C.P. Stark, and P.A. Allen (1997): Sediment flux from a mountain belt derived by landslide mapping. Geology, 25(3): 231–234.

Hutchinson, L.N., and R. Bhandari (1971): Undrained loading – a fundamental mechanism of mudflows and other mass movements. Geotechniques, 21.

Iverson, R.M. (1997): The physics of debris flows. Rev. Geophys., 35/3, 245–296.

Jan, C.-D., and H.W. Shen (1997): Review dynamic modelling of debris flows. In: Recent developments on debris flows, edited by A. Armanini and M. Michiue, Springer, LNES 64, 93–116.

Kirkby, M.J. (1998): Landscape modelling from regional to continental scales. In this volume.

Kröner, D. (1997): Numerical schemes for conservation laws. Wiley, Chichester.

Ohmori, H. (1992): Morphological characterisitcs of the scar created by large–scale rapid mass movement. Japan. Geomorph. Union Trans., 13: 185–202.

Rinaldo, A., I. Rodriguez–Iturbe, R. Rigon, and R.L. Bras (1993): Self–organized fractal river networks. Phys. Rev. Lett., 70: 822–825.

Sapozhnikov, V.B., and E. Foufoula–Georgiou (1996): Do the current landscape evolution models show self–organized criticality? Water Resour. Res., 32(4): 1109–1112.

Sapozhnikov, V.B., and E. Foufoula–Georgiou (1997): Experimental evidence of dynamic scaling and indications of self–organized criticality in braided rivers. Water Resour. Res., 33(8): 1983–1991.

Somfai, E., A. Czirok, and T. Vicsek (1994): Power–law distribution of landslides in an experiment on the erosion of a granular pile, J. Phys.: Math Gen, 27: L757–L763.

Sugai, T., H. Ohmori, and M. Hirano (1994): Rock control on the magnitude–frequency distribution of landslides. Trans. Japan. Geomorphol. Union, 15: 233–251.

Takayasu, H., and H. Inaoka (1992): New type of self–organized criticality in a model of erosion. Phys. Rev. Lett., 68: 966–969.

Yokoi, Y., J. R. Carr, and R. J. Watters (1995): Fractal character of landslides, Envir. & Engin. Geoscience, 1: 75–81.

Tectonic Predesign in Geomorphology

R. Hantke[1] and A. E. Scheidegger[2]

[1] Institute of Geology, ETH Zürich, Switzerland
[2] Department of Geophysics, TU Wien, Austria

Abstract. The evolution of landscapes is governed by a few simple principles which can be viewed in the context of complexity–theory: landscapes can be regarded as open nonlinear systems in which tectonic processes furnish the "input" and denudational processes the "output" ("Antagonism Principle"). Stationary states in a landscape correspond to self–structured order in a complex system; they exist in a limited range of space and time only, they are "selected" for some finite duration in a limited region ("Selection Principle"). Geomorphology has mostly been concerned with the "inner" workings of a landscape–system; – i.e. mainly slow process–response phenomena. However, the tectonic input ("Principle of Tectonic Predesign") strongly influences the genesis of many geomorphic landscape features such as drainage systems, the shape of valleys, incised meanders, glacial forms, mass movements and other features: some common contentions (e.g. that water causes V–shaped, ice U–shaped valleys) are shown to be in need of modification.

Introduction

Phenomenologically, the evolution of landscapes is governed by a few simple principles. These have recently been summarized by Scheidegger (1987, 1991): they are the Principle of Antagonism, the Selection Principle and the Principle of Tectonic Predesign.

These landscape principles can be viewed in the context of complexity theory: landscapes can be regarded as open, nonlinear, complex systems, in which tectonic processes furnish the "input" and denudational processes the "output" ("Principle of Antagonism", Scheidegger 1979b). The "input" is statistically well–structured and systematic, because plate–tectonic processes reach over whole continents. On the other hand, the "output" occurs (phenomenologically) in a random–fashion, because the meteorologic and fluid–dynamic processes have generally temporal and spatial correlation–ranges which are much smaller than the features concerned. Thus, the "tectonic" and the "exogenic" aspects in a geomorphic feature can be inferred from morphometric statistics.

Stationary states in a landscape correspond to self–structured order in a complex system (otherwise they would not be of any duration and the landscape features in question could not be described), determined by a strange attractor at the edge of chaos: it seems to be a natural law that every observable region in an open, nonlinear, complex system is self–structured on a strange attractor

(Scheidegger 1996) which is characterized temporally by 1/f–noise and spacially by fractality, the latter implying a power–law for the number of spacial subsets, e.g. the Gutenberg–Richter (1949) law for the magnitude vs. frequency of the earthquakes in any given region. Stationary states, corresponding to a process–response mechanism, exist in a limited range of space and time only, they are "selected" for some finite duration in a limited region (this is the expression of Gerber's (1969) "Selection Principle" in complexity–theory terminology). Thus a landscape is not unifractal (Evans and McClean 1995) – neither in space nor in time; already Gutenberg and Richter (1949) assumed the existence of a "maximum" magnitude in their magnitude–frequency relationship, and the linearity does not extend to very low magnitudes, either.

Finally, the "Principle of Tectonic Predesign" states that many geomorphic features, such as drainage systems, the shape of valleys (V–, U–form), incised meanders, glacial forms and mass movements are controlled by ongoing tectonic processes; this is not quite the same as the idea of pre–existing tectonic structures being reflected in structural landforms associated with folds etc. (Twidale 1971). The orientation structure of the neotectonic stress field can be determined at any locality by in–situ measurements, from well–breakout observations, from earthquake fault–plane solutions, and, most easily, by making joint orientation measurements and evaluating them statistically (Kohlbeck and Scheidegger 1977, 1986); the bisectrices of the two conjugate joint sets agree with the local principal stress directions, world wide (Scheidegger 1995). The fact that the local joint strikes agree with the local trends of many geomorphic features (noted by the present authors, but also by Ahnert 1996: 289) would indicate that the latter are (neo–)tectonically predesigned.

Valley Features

General Remarks

Valleys usually conform strictly to tectonic structures, they follow hollows, split–up anticlines, heads of layers, fronts of nappes and they break through mountain chains along tectonic faults (Hantke 1991: 19). We shall present a few characteristic cases of this type.

Drainage Patterns

The drainage patterns in areas of medium to high relief have often been considered as antecedent to the orogeny. Thus, Staub (1934, see also Ahnert 1996: 254) postulated that the drainage pattern of the rivers in the central plain of Switzerland is the carbon–copy of an antecedent purely "erosional" Miocene drainage pattern in a plane. However, Scheidegger (1979a,b) has shown that the distribution of the directions of the links agrees with that of the neotectonic joints (Fig. 1) so that it must be held that the drainage orientation pattern in Switzerland is completely determined by neotectonic processes. Similar cases have been reported from all over the world; Scheidegger and Ai (1986) have given a corresponding review.

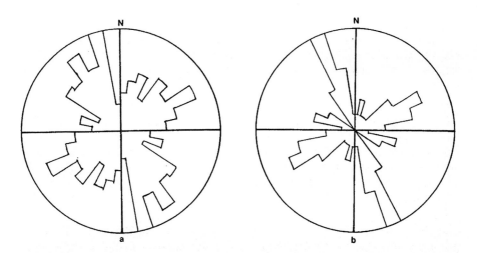

Fig. 1. Swiss plain: trends of valley directions (a) compared with strikes of joint orientations (b). After Scheidegger (1979a).

Widening of Valleys

The shape of fluvial valleys is generally ascribed to intrinsic fluvial processes. In this, it is usually maintained that the "natural" or "original" (youth) form of a valley is V–shaped (Davis 1924: 90, Holmes 1944: 153, Holmes 1965: 479, Ahnert 1996: 277). As they deepen, valleys become widened which fact is ascribed to the scouring and steepening of the channel sides by the river itself, to rainwash and gullying, and to soil creep (Holmes 1944: 160ff.). However, there is now overwhelming evidence that the erosive action is insufficient for the creation of the valley profiles: tectonic stress–induced faulting and jointing must be responsible for the latter. Hantke (1991: 41) presents several examples bearing this out in the valleys of the Luetschine (Bernese Alps, Switzerland, Fig. 2) and of the Maggia (Ticino, Switzerland); joint– and valley–orientations usually coincide.

Transverse Valleys

In many locations, such as in the Jura mountains or in Alpine foothills of Switzerland, it is the established view that valleys cutting transversally through mountain chains (often in a gorge, German Klus, French cluse; Fig. 3) are antecedent to the genesis of these chains (Brunner 1909: 163 regarding the "cluses" in the Jura Mountains, Labhart 1991 regarding the Aare Gorge in the Alpine foothills, Ahnert 1996: 255 for a general discussion of the problem). However, Hantke (1991: 257) as well as Young and McDougall (1993) note that the river action in such gorges is much too small to keep up with the orogeny. Heuberger (1975) found that the downward incision into solid rock caused by the outflow from a lake formed some 8000 years ago by the Koefels (Tyrol) slide was only a few meters since the slide occurred, i.e. less than 0.5 mm/a – much too little

Fig. 2. Alluvial plain of the Luetschine, Lauterbrunnen Valley, Ct. Berne.

in order to keep up with the Alpine uplift rate of several mm/a. Indeed, the hydraulic bottom shearing stress in a river is of the order of 100 kPa (Magilligan 1992), whereas the shearing strength of rocks is of the order of 10 MPa (Scheidegger 1982:188–189) so that a river can never "saw" through a (rising) mountain. This is only possible if there is some tectonic shear zone (or such like) which the river merely has to ream out rather than cut through it. For the Aare gorge, Müller (1938) had already found joint–systems which are paralleling the latter, an observation which has been further tested by the present authors (Hantke and Scheidegger 1993). Thus, the gorges must owe their genesis to the existence of tectonic shear zones, buckling or transverse faults which are then used by the river to make its way.

Incised Meanders

Meanders in river courses are on occasion stabilized in deep gorges. Famous examples are the Aare loop around Berne, Switzerland, the Wear loop around Durham, England, or the "Goosenecks" of the San Juan River in Utah, U.S.A., amongst many others.

Generally, it has been assumed that such incised meanders are due to the rejuvenation of a landscape with only a thin cover of easily eroded deposits; the deepening channel is etched into the underlying (solid) rocks with the original winding course being preserved (Holmes 1944:197, Holmes 1965). Ahnert (1996:220) suggests that the development of such meanders is aided by the presence of vertical (not necessarily tectonic) structures in the underlying rocks.

Fig. 3. The Klus of Moutier in the Swiss Jura.

However, there may be a significant neotectonic input present: thus, Hantke (1991: 170) postulates that the loop around Berne is determined by shear faults. On a smaller scale, the "meandering" of gullies by zig–zagging along conjugate joints in Karoo volcanics of Southern Africa is patently evident (Fig. 4).

Fig. 4. Zig–zagging gulleys in the Karoo Formation near Umtata, Transkei, Southern Africa.

Longitudinal River Profiles

It is well known that tectonic events can significantly influence the longitudinal profile of a river. Thus, for instance, it is obvious that the cataracts of the Nile have been caused by the intrusion of igneous materials into the surrounding country rock (Said 1962): e.g. near Aswan by the intrusion of Aswan Granite into Nubian Sandstone (Fig. 5). This qualitative statement can be quantified by a numerical comparison of actual river profiles with theoretical ones that would be the result of purely fluvial processes. The "normal" geomorphic longitudinal river profiles are exponential or logarithmic (Scheidegger 1991: 208–209); however, there are commonly "knick–points" present whose initiation has often been ascribed to some geomorphic instability. Nevertheless, the classical view seems to be that no equilibrium is possible around a knick–point and that the latter must therefore wander upstream and become obliterated (Scheidegger 1991: 209). Thus, stable knick–points cannot be due to the intrinsic river action alone.

Fig. 5. First cataract of the Nile River near Aswan, Egypt.

In this vein, a very detailed numerical analysis of river profiles in the Tarai district of West Bengal at the base of the foothills of the Eastern Himalayas has been made by Maiti (1980:60, 1991:114ff., see also Mukhopadhyay 1982) who showed that none of the theoretical geomorphic profiles can be made to fit the actual data. Knick points can be identified with the traces of upthrust planes; moreover, the head–water regions are over–steepened which can only be explained by the continuing uplift of the Himalayas (Fig. 6 as an example). Moreover, the size grading found in natural rivers (e.g. Standing Stone Creek, Pa.) cannot be produced by sorting and abrasion; the sizes must be supplied by the downstream tributaries which, again has to have a tectonic origin (Pizzuto 1992).

Glacial Features

General Remarks

Next, we turn to features whose form has generally been ascribed to glacial action. This refers to the supposed U–shape and overdeepening of glacial valleys, piedmont lakes and fiords, as well as to the genesis of roches moutonnees. The orientation structure of the glacial "drainage" patterns has, of course, a similar origin as that of fluvial drainage (Scheidegger and Ai 1986).

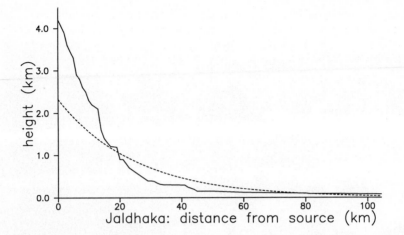

Fig. 6. Longitudinal profiles (actual profile: solid line; and best–fitting theoretical–exponential profile: broken line) of the Jaldhaka River, West Bengal, India. Ordinate: elevation (km) above sea level; abscissa: distance (km) of thalweg downstream from the source. Note knick points and oversteepened head water region in the actual profile. Based on data from Mukhopadhyay (1982: 57).

Glacial U versus Fluvial V Valleys

The contention that glacial valleys are supposed to be U–shaped, fluvial valleys V–shaped goes back at least to Davis (1909), if not further. Since that time, it has been repeated in many textbooks on physical geology or on geomorphology (e.g. Holmes 1944: 204–252, Machatschek 1952: 128–129), although some multifactorial elements have occasionally been introduced (Ahnert 1996: 337). Undoubtedly, the erosion by flowing ice has had a pronounced effect: some ice–scoured valleys in Scandinavia do indeed have a U–shape; however, there are many V–shaped glacial valleys in Switzerland (Hantke 1991: 44). Furthermore, in flat–lying carbonate rocks the valley sides break off almost vertically and the debris at the valley floor simulate a U–shape (Fig. 7). In such cases, the valley shape is tectonically predesigned since purely glacial–erosive processes cannot explain the appearance (Gerber 1945): the cause of the morphology are "tectonic lineaments", i.e. faults and joints (see also Sonder 1938). Thus, the erosive action of flowing ice has often been overestimated, and it is not tenable to assume that all V– shaped valleys are of fluvial, all U–shaped valleys of glacial origin.

Overdeepening by Glacial Erosion

Another point is the overdeepening of valleys, supposedly caused by glacial or fluvial action (eg. Holmes 1944: 224, Ahnert 1996: 337). The infilling of these valleys by fluvial action can be excluded; the fill is of glacial origin (Schluechter 1987). In fact, the assumption that glacial or fluvial erosive action is at the root of

the excavation of valleys, led to a disaster on 24 July 1908 during the building of
the Lötschbergtunnel in the Bernese Oberland: water–logged gravel masses were
encountered 160 m underneath the present valley floor; the resulting flood was
fatal to a shift of workers in the tunnel (Schluechter 1983, Labhart 1991:126). A
similar surprise was encountered when seismic investigations (Pfiffner et al. 1997)
showed that the Rhone glacier near Martigny had "overexcavated" the valley to
450m below sea level – 900m below the present–day valley floor. The existence
of loose gravel at such depths is unconceivable in the light of the theory that
a glacier or a river had excavated the valley. Furthermore, the tributary creeks
of the "glacial" piedmont lakes are not ice–edge features; their orientation fits
the neo–tectonic stress/joint pattern (Hantke and Scheidegger 1997). Thus, the
primary lineaments, hollows and clefts owe their existence not to erosion, but
rather to various tectonic processes.

The same holds true for the "overdeepening" of piedmont lakes (Fig. 8, Hantke 1991:117). Thus, it is not the glacial action alone which has caused such
features; rather, there are tectonic processes involved.

"Glacial Overdeepening" has also been held responsible for the genesis of
fiords; in fact, the idea that glaciation has played a major role in the formation
of fiords seems generally accepted (Holtedahl 1967). A completely opposite view
has been put forward by Gregory (1913 and in many later publications) who
maintained that fiords are of tectonic origin and that the glacial influence was
negligible during their genesis. Holmes (1944:224) combined the two views and
admitted that, in addition to glacial action, an "appropriate structure" of the
respective region may have been involved. He also noted that in plan, fiords
have everywhere a rectilinear pattern, which he claims to be "clearly determined" by belts of structural weakness, such as synclines of weak sediments or
schists enclosed in crystalline rocks. We have tried to make a contribution to
this controversy by studying a typical fiord–coast: that of Spitzbergen (Fig. 9).
Orientation data of 249 joints were taken (by A.E.S.) at Barentsburg, Longyearbyen,Magdalenefiord and Ny Alesund (B,L,M,N in Fig. 9) and compared with
the trend of the fiords, indicated by heavy lines in Fig. 9. There were basically two
conjugate joint sets whose strikes (statistical maxima – calculated by the method
of Kohlbeck and Scheidegger 1977 – at N86°E and N179°E) are close to the two
conjugate trends of the fiords (statistical maxima at N77°E and N161°E): the
E–W striking trends are only 9° apart, which, within the error limits, can be
regarded as a coincidence. The corresponding strike (trend) roses are shown in
Fig. 10. Whilst such observations cannot, of course, lead to any final conclusions,
they do point towards an identical origin of joints and fiords; inasmuch as joints
are known to be generally of a neotectonic origin, the same is likely to be true
for the fiords as well: their rectilinear patterns noted by Holmes (l.c.) would then
correspond to the conjugate shear lines of the neotectonic stress field.

Roches Moutonnees

Roches moutonnees have been considered as the result of nonuniform, unstable
erosion on a hard glacier bed. Roches moutonnees show a smoothly abraded

Fig. 7. The Obere Ochsenkar in the Hochkoenig area of Salzburg Province, Austria, which creates the impression of a U–shaped valley with its lateral debris slopes.

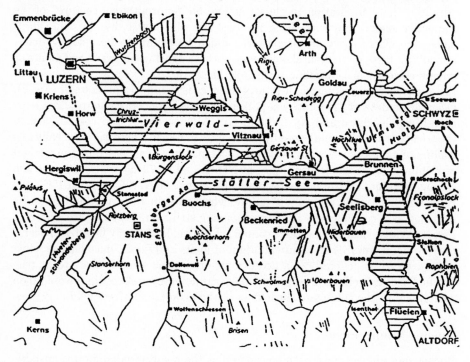

Fig. 8. Tectonic predesign of the Lake of Lucerne Basin. After Hantke 1991, p. 117.

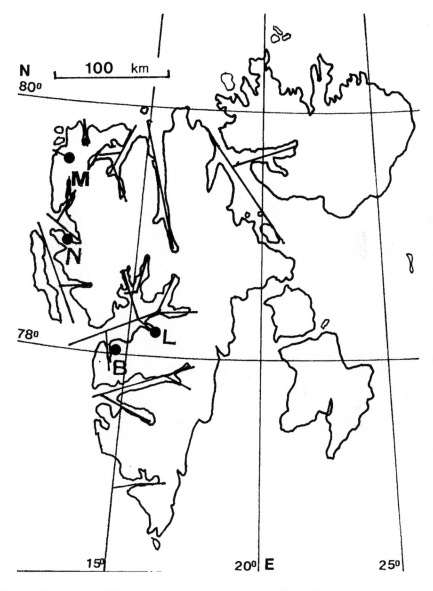

Fig. 9. Sketch–map of Spitsbergen with the locations of the joint–orientation measure-ments (B: Barentsburg; L: Longyearbyen; M: Magdalenefiord; N: Ny Alesund) marked by dots and the trends of the fiords marked by heavy lines. The statistical weight of each fiord–trend was taken according to its length.

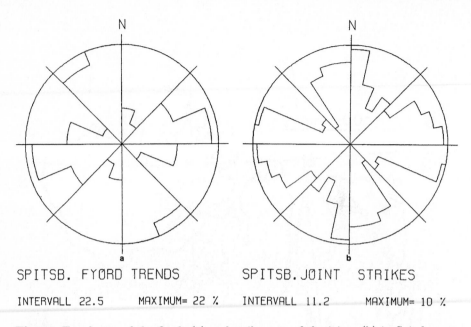

SPITSB. FYØRD TRENDS   SPITSB. JØINT   STRIKES

INTERVALL 22.5   MAXIMUM= 22 %   INTERVALL 11.2   MAXIMUM= 10 %

Fig. 10. Trend–rose of the fiords (a) and strike–rose of the joints (b) in Spitsbergen: the respective maxima are very close to each other.

slope on the side up which the ice advanced; the lee side falls more steeply. Even here, a step–like series of crags and ledges due to the plucking out of joint blocks (Fig. 11) is commonly observed (Holmes 1944: 217). Ahnert (1996: 290) notes that joints can indeed delimit the faces of any rock–mound. However, there is more to it: since joints have been found to be connected with the tectonic stress field, it is clear that tectonic processes play a major role in the predesign of such features; they are not caused by an instability of the external erosion (Hantke 1991: 51).

Mass Movements on Slopes

Mass movements on slopes are part and parcel of the normal landscape develop-ment cycle; the ongoing uplift is compensated by such movements which leads to a quasi–stationary state by a process–response mechanism (Carson and Kirkby 1972) "at the edge of chaos". Nevertheless, it has been found that in many mass movements there are also aspects that are influenced by tectonic processes.

Regarding all sorts of mass movements (slow slumps, debris flows, landslides; for a general qualitative characterization see e.g. Dikau et al. 1996: 1–12) one can make the following general statements:

The mass movements (i) do occur in spurts, they are part of the normal landscape development (their incidence is a stochastic time series with a fractal structure) and (ii) their direction is predesigned by the neo–tectonic stress–field;

Fig. 11. Jointed roche moutonnee in the Hochkoenig area of Salzburg Province, Austria.

the latter also generates the orientation of the joints, hence the direction of the joints and the slide–motion directions are correlated.

Thus, whilst the immediate triggering of mass movements on slopes is undoubtedly due to external (meteorologic or seismic) processes (Iverson et al. 1997), their location and orientation is influenced by tectonic processes. In this connection, orientation studies of the directions of mass movements in relation to the prevailing neotectonic stress field (generally established by joint orientation measurements) have been reported from many parts of the world (see Scheidegger and Ai 1986 for a review): the direction of the mass movements is naturally in the direction down the slope of the valleys. If the latter are natural, they are parallel to the prevailing joints which are shearing features of the neotectonic stress field. As an example, we show (Fig. 12) the coincidence of the orientation of the motion vectors (N131°E±4°) with that of one of the preferred joint sets (strike N127°E±8°) found in the Graukogel slide near Bad Gastein, Salzburg Province, Austria (Hauswirth and Scheidegger 1980). However, this is not true in the case of artificial cuts: in that case, slides occur mainly on faces running at right angles to the maximum neotectonic compression direction. Evidently, the stability of the object is reduced in this case and slides occur more frequently than if the cut runs parallel to the maximum compression (Ai and Scheidegger 1984).

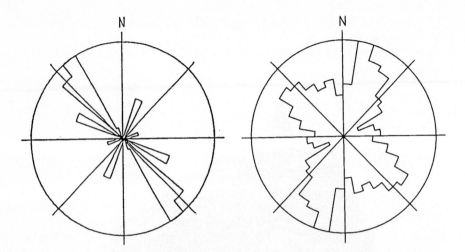

Fig. 12. Rose diagram of displacement vectors (left) and rose (right) of joint strike–directions in the mass–movement area of Badgastein in Salzburg Province, Austria: The displacement vectors coincide with one of the principal joint–strike orientations; they are also normal to the valley orientations (not shown here). Modified after Hauswirth and Scheidegger 1980.

Conclusions

The foregoing discussions have shown that the basic design of particular features, whilst existing as quasi–stationary "forms" at the edge of chaos in a complex system, is often influenced by tectonic processes, not by process–response behavior alone. This is especially evident in connection with the view that rivers cause V–shaped, glaciers U–shaped valleys: neotectonic influences have been shown to be important. Other instances, such as that of incised meanders and that of the orientation structure of drainage basins and fiords, amongst others, have also been mentioned: the influence of tectonic processes (and not only of preexisting tectonic structures) in geomorphology is profound: the qualitative multifactorial approach to the genesis–problem of landscape features which has slowly been gaining entry into geomorphology (Ahnert 1996 as well as other authors), needs to be extended and, most of all, quantified by the introduction of more statistical and dynamical–analytical considerations.

References

Ahnert, F. (1996): Einführung in die Geomorphologie. UTB Verlag Ulmer, Stuttgart.
Ai, N.S., and A.E. Scheidegger (1984): On the connection between the neotectonic stress field and catastrophic landslides. Proc. Internat. Geol. Congr. 27th., Moscow K06: 180–184.
Brunner, H. (1909): Die Schweiz. Attinger, Neuenburg.

Carson, M.A., and M.J. Kirkby (1972): Hillslope Form and Process. University Press, Cambridge.

Davis, W.M. (1909): Geographical Essays. Ginn, Boston (Reprint Dover New York, 1954).

Davis, W.M. (1924): Die erklärende Beschreibung der Landformen, 2nd ed., Teubner, Leipzig, 565pp.

Dikau, R., D. Brunsden, L. Schrott, and M.-L. Ibsen (eds.) (1996): Landslide Recognition: Identification, Movement and Causes. Wiley, New York.

Evans, I.S., and C.J. McClean (1995): The land surface is not unifractal: variograms, cirque scale and allometry. Z. Geomorph. Suppl. 101: 127–147.

Gerber, E. (1945): Lage und Gliederung des Lauterbrunnentales und seiner Fortsetzung bis zum Brienzersee. Mitt. Aarg. Naturf. Gesellsch. 22: 165–184.

Gerber, E.(1969): Bildung und Formen von Gratgipfeln und Felswaenden in den Alpen. Z. Geomorph. Suppl. 8: 94–118.

Gregory, J.W. (1913): The Nature and Origin of Fjords. Murray, London.

Gutenberg, B., and C.F. Richter (1949): Seismicity of the Earth and Associated Phenomena. Princeton University Press, Princeton.

Hantke, R. (1991): Landschaftsgeschichte der Schweiz und ihrer Nachbargebiete. Ott, Thun.

Hantke, R., and A.E. Scheidegger (1993): Zur Genese der Aareschlucht. Geographica Helvetica 48/3: 120–124.

Hantke, R., and A.E. Scheidegger (1997): Zur Morphogenese der Zuerichseetalung. Vierteljahrsschrift der Naturf. Gesellsch. Zuerich 142/3: 89–95.

Hauswirth, E.K., and A.E. Scheidegger (1980): Tektonische Vorzeichnung der Hangbewegungen im Raume von Badgastein. Interpraevent 1980, 1: 159–178.

Heuberger, H. (1975): Das Oetztal. Bergstürze und Gletscherstände, kulturgeographiische Gliederung. Innsbrucker Geographische Studien 2: 213–249.

Holmes, A. (1944): Principles of Physical Geology, 1st ed., Nelson, London.

Holmes, A. (1965): Principles of Physical Geology, 2nd ed., Nelson, London.

Holtedahl, H. (1967): Notes on the formation of fjords and fjord–valleys. Geografiska Annaler 49A/1–2: 188–203.

Iverson R.M., M.E. Reid, and R.G. LaHusen (1997): Debris–flow mobilization from landslides. Ann. Rev. Earth Planet. Sci 25: 85–138.

Kohlbeck, F.K., and A.E. Scheidegger (1977): On the theory of the evaluation of joint orientation measurements. Rock Mechanics 9: 9–25.

Kohlbeck, F.K., and A.E. Scheidegger (1986): The power of parametric orientation statistics in the earth sciences. Mitt. Österr. Geol. Ges. 78: 251–265.

Labhart, T.P. (1991): Geologie der Schweiz. Ott, Thun.

Machatschek, F. (1952): Geomorphologie, 5th ed., Teubner, Leipzig.

Magilligan, F.J. (1992): Thresholds and spatial variability of flood power during extreme floods. Geomorphology 5: 373–390.

Maiti, G.S. (1980): Quantitative analysis of the Jaldhaka Basin. Indian J. Landscape Systems and Ecological Studies 3/1–2: 58–66.

Maiti, G.S. (1991): An Analytical Study of the Landforms and Land Uses of the Tarai Area (the Eastern Himalayas), West Bengal. Doctoral Thesis, Dept. Geography, Calcutta University, Calcutta.

Müller, F. (1938): Das Gebiet der Aareschlucht, p. 42–47 In: Geologie der Engelhoerner, der Aareschlucht und der Kalkkeile bei Innertkirchen (Berner Oberland), Beitr. zur Geolog. Karte der Schweiz N.F. 74: I–X: 1–55.

Mukhopadhyay, S.C. (1982): The Tista Basin, a Study in Geomorphology. Bagchi & Co., Calcutta.

Pfiffner, O.A., et al. (ed.) (1997): Deep Structure of the Swiss Alps: results of NRP 20. Birkhäuser, Basel.

Pizzuto, J. (1992): The morphology of graded rivers: a network perspective. Geomorphology 5: 457–474.

Said, R. (1962): The Geology of Egypt. Elsevier, Amsterdam.

Scheidegger, A.E. (1979a): Orientationsstruktur der Talanlagen in der Schweiz. Geographica Helvetica 34/1: 9–15.

Scheidegger, A.E (1979b): The principle of antagonism in the Earth's evolution. Tectonophysics 55: T7–T10.

Scheidegger, A.E. (1982): Principles of Geodynamics, 3rd ed., Springer, Berlin.

Scheidegger, A.E. (1987): The fundamental principles of landscape evolution. Catena Suppl. 10: 199–210.

Scheidegger, A.E. (1991): Theoretical Geomorphology, 3rd ed., Springer, Berlin.

Scheidegger, A.E. (1995): Geojoints and geostresses. In: Mechanics of Jointed and Faulted Rock, Proc. 2nd. Internat. Conf. on the Mechanics of Jointed and Faulted Rocks, ed. by H.P. Rossmanith, Balkema, Rotterdam: 1–35.

Scheidegger, A.E. (1996): Ordnung am Rande des Chaos: ein neues Naturgesetz. Österr. Z. f. Vermessung und Geoinformation 84/1: 69–74.

Scheidegger, A.E., and N.S. Ai (1986): Tectonic processes and geomorphological design. Tectonophysics, 126: 285–300.

Schluechter, C. (1983): Die Bedeutung der angewandten Quartärgeologie fuer die eiszeitgeologische Forschung in der Schweiz. Physische Geographie (Zuerich) 11: 59–72.

Schluechter, C. (1987): Talgenese im Quartär – eine Standortbestimmung. Geogr. Helv. 42/2: 109–115.

Sonder, R.A. (1938): Die Lineamenttektonik und ihre Probleme. Ecl. Geol. Helv. 31: 199–238.

Staub, R. (1934): Grundzüge und Probleme alpiner Morphologie, Denkschr. Schweiz. Naturforsch. Gesellsch. 69/1: 1–183.

Twidale, C.R. (1971): Structural Landforms. MIT Press, Cambridge MA.

Young, R., and I. McDougall (1993): Long–term landscape evolution: Early Miocene and modern rivers in Southern New South Wales, Australia. J. Geol. 101: 39–49.

Modelling Changes in Terrain Resistance as a Component of Landform Evolution in Unstable Hill Country

M. J. Crozier and N. J. Preston

Department of Geography
Research School of Earth Sciences
Victoria University of Wellington, New Zealand

Abstract. Accurate modelling of landform evolution in unstable terrain requires some means of determining the frequency and magnitude of landslide occurrence. In many areas, this can be achieved with reference to the potency of the triggering regime and the susceptibility of the terrain. Terrain susceptibility is controlled by inherent geomechanical resistance and topographic conditions which filter the effect of the triggering agent. For a given terrain, susceptibility can be conveniently represented by establishing the triggering threshold required for the occurrence of landslides. Problems are identified in use of thresholds developed from historical and contemporary measurements of landslide activity to predict or postdict landslide activity. Even if all other external controlling factors remain constant, there appear to be event–related temporal shifts in terrain resistance which change the level of the triggering threshold and the occurrence of landslides over time. Mechanisms responsible for this change are proposed and the results of field investigation into geomechanical limiting equilibrium conditions are produced to demonstrate the effects of these mechanisms and the changes in resistance they produce. A conceptual model is presented to place event–related changes in susceptibility into the context of a major phase of landslide–induced soil erosion which has affected New Zealand hill country since European deforestation.

Introduction

Landslides represent an important mechanism in landform evolution. In many areas of the world and at different periods this mechanism dominates both the reduction of mass and change in landform geometry (Crozier et al. 1992). The extent to which this mechanism can operate is a function of terrain resistance, i.e. how readily the terrain can resist the physical destabilising stresses imposed upon it. Thus, in order to understand and predict the rate and manner of landform evolution, a knowledge of how terrain resistance changes through time is required.

Landslides are also significant from an environmental perspective because they represent an easily identifiable, short–lived, self–annihilating adjustment to external perturbation. Major landslide activity thus often signals a change or disturbance in ambient environmental conditions. Landslides can, therefore, be a significant indicator of environmental change, responsive to change in climate, vegetation, hydrologic, tectonic, and human–induced disturbances. Of more immediate concern is their importance as a natural hazard (Glade and Crozier 1996). The degree of response to disturbance, the level of hazard, and rate of geomorphic change are all usefully represented by measures of frequency and magnitude of landslide occurrence (Crozier 1996).

Fundamental Concepts

It is consistent with geomechanical theory that a landslide will occur when an externally imposed stress (trigger) causes the shear strength/shear stress ratio (factor–of–safety) of a slope to drop below 1.0 (Selby 1993). In any given terrain, the frequency with which this will happen (probability of occurrence of landslides) depends on two fundamental conditions:

1. The potency of the triggering regime, i.e. the energy within the triggering regime measured, in any given terrain, by its ability to exceed a landslide triggering threshold in terms of frequency, magnitude, and duration.
2. The susceptibility of the terrain.

Susceptibility, in turn, is made up of inherent resistance (measured by the factor–of–safety and its inherent ability to be reduced to unity) and the local triggering enhancement conditions (ambient filters). Ambient filters are site factors, such as: upslope catchment area, vegetation cover, and topographic form characteristics (e.g. plan curvature) which serve to enhance or filter the effect of the triggering agent. By integrating these factors, susceptibility is conveniently represented and measured by the triggering agent threshold required to produce landslides, in a given terrain (Crozier 1989). It has become common practice to identify thresholds either theoretically, or empirically. Having established a threshold, the behaviour of the triggering agent regime can then be analysed to determine the probability of landslide occurrence from the probability of threshold excedence.

Successful application of thresholds for predicting the frequency and magnitude of future landslide activity depends on a number of factors. These include the accuracy with which the threshold was able to represent limiting conditions when it was originally developed, the homogeneity of the terrain and, because of temporal changes in potency and susceptibility, the length of the period to which the threshold is applied. Accuracy of the thresholds must be known in order to place confidence limits on predictions. Some attempts to construct thresholds yield a broad threshold zone rather than a well–defined limiting condition. Threshold zones can be characterized by their upper and lower bounds. The lower bound has been referred to as the minimum threshold, below which no landsliding occurs, and the upper bound represents the maximum threshold (envelope), above which slipping is always recorded (Crozier 1989).

Another important consideration, discussed by Crozier (1996), is the positive relationship between the magnitude of the triggering agent and the magnitude of landslide response, as measured by such parameters as, volume of material displaced, area affected, and number of landslides produced. The implication is that discussion of thresholds should not only indicate their accuracy, i.e. whether they refer to 'maximum' or 'minimum' limits but should also be qualified by the magnitude of response they are representing. Thus it is possible to represent the work of landsliding in a similar way to frequency/magnitude analysis used in fluvial hydrology.

When assessing the frequency and magnitude of landslide activity with respect to landform evolution, it is important to distinguish between the concepts of potency and susceptibility because they can change independently, thereby affecting the probability of occurrence of landslide activity. The influence of spatial differences in both potency and susceptibility in controlling the frequency of occurrence from place to place is well appreciated. Less well understood is the role of temporal variation in potency and susceptibility in affecting the frequency of occurrence of landslides, in any one place. It is the temporal variation in susceptibility which forms the subject of this paper. Understanding and incorporating temporal changes of both potency and susceptibility is essential for successful modelling of process–based landform evolution.

Thresholds and Landform Evolution in Unstable Terrain of New Zealand

The extent to which threshold–based frequency/magnitude analysis of landslide occurrence can be used to predict or postdict landform behaviour clearly depends, not only on the accuracy of the analysis, but also on the stability of the land/environment system in question. Rapidly changing systems have the potential to affect both the potency and susceptibility of landslide generating conditions. New Zealand is by world measures, a high energy, tectonically active area which, in addition, has been highly affected by human activity within a short period of time. Within the space of less than 1000 years the country has experienced the arrival of Polynesians and, within the last 200 years, colonisation by Europeans. Although natural disturbances such as volcanic activity, drought and fire have produced transient changes to New Zealand's indigenous podocarp/hardwood forests in the past, the first major change resulted from Polynesian–induced burning about 500 years ago. However, pollen and sedimentation analysis indicates that Polynesian disturbance resulted in a localized change from forest to fern with little detectable increase in erosion rates (Wilmshurst 1997). There is, however, an on–going debate on the effect of Polynesian deforestation on erosion rates. By contrast, within the first two or three decades of European settlement, most of New Zealand hill country experienced widespread conversion from forest, scrub, or fern to introduced pasture and the consequent initiation of a dramatic phase of soil loss by landsliding. We refer to this major increase in erosional activity as the 'Post–European Settlement Regolith Stripping Phase' because it takes place through the episodic occurrence of rainfall–triggered landslides which individually remove most of the regolith from the bedrock surface and redistribute it or export it from the system (Fig. 1).

Fig. 1. Rainfall–triggered, shallow landslides indicating extensive removal of regolith, Kaweka Range, North Island, New Zealand.

Given this degree of environmental and slope instability, the critical question is: to what extent is it possible, from an historical analysis of the landslide/climate relationship, to establish a regional threshold and thus reliably predict the probability of landslide occurrence and rate of landform evolution? The large number of landslide episodes that have been recorded in New Zealand in historical times (see Glade and Crozier (in press) for a full bibliography) certainly makes it relatively easy to obtain a large database for threshold development. However, there is mounting evidence that this problem is compounded by the existence of temporally unstable or shifting thresholds that are an intrinsic component of unstable land systems (Page et al. 1994). This problem would appear to affect all unstable systems to a greater or lesser extent and will therefore need to be factored into process–based landform evolution models. From work carried out in New Zealand, we propose that changes in the frequency of occurrence of landslides and the degree to which they will affect landform evolution is not only affected by changes in potency but also by event–induced intrinsic shifts in inherent susceptibility. The establishment of thresholds for different time periods and an assessment of the rate and manner with which they change through time are obviously key factors in successfully modelling landform evolution.

The Terrain Event–Resistance Model

Description

The way in which we think inherent susceptibility changes in landslide–affected hill country terrain is shown in the conceptual model (Fig. 2). The model represents changes in susceptibility by indicating the notional rainfall threshold required to produce a given magnitude of landslide activity, at different stages in the regolith stripping process. The model is specific for a number of reasons: the primary destabilising perturbation is large scale deforestation, there are distinct hydraulic and strength differences between regolith and bedrock, individual landslides occur close to the bedrock surface, and also because it assumes that most regolith will eventually be removed from steep hill country catchments. The extent to which these attributes constrain application of the model beyond the New Zealand setting is uncertain. However, it is evident that many of these attributes are common, in some form or another, to many landslide–affected areas around the world.

In a broad sense, the concept represents a relaxation model applied to thresholds rather than form, indicating trends between two metastable but different states, the 'forest phase' and the 'bedrock phase'. A distinctive feature of the model is the acknowledgment that relaxation can be intrinsically interrupted. The interruption occurs during the overall process of exhaustion of the available material (regolith). Availability of material is seen as the dominant factor constraining landslide activity; in the 'primary adjustment' phase this involves in situ (undisturbed) regolith, whereas in the 'secondary adjustment' phase the accumulation of depositional (disturbed) material, in certain locations, temporarily counteracts the overall trend of decreasing susceptibility.

Mechanisms and Justification

The role of deforestation as a major destabilising influence has been well documented elsewhere both in terms of mechanisms (Greenway 1987) and effects (O'Loughlin and Pearce 1976). In New Zealand, Selby (1976) and Crozier and Pillans (1991) have shown susceptibility to landsliding has on average increased by about three times as a result of deforestation and conversion to pasture. In the Lake Tutira catchment (Hawke's Bay) which is typical of much of New Zealand's landslide–prone hill country, Page and Trustrum (1997) have established from lake core analysis that soil removal from catchment hillslopes in the post–European settlement period (from A.D.1878) has been 5–6 times the rate under fern and scrub (560 yr B.P. to AD 1878) and 8–17 times the rate under indigenous forest (1850–560 yr B.P.).

The degree to which regolith stripping has advanced since deforestation has also been well–established from field and airphoto surveys in different parts of the country (Crozier et al. 1980). The extent to which this regolith stripping process itself affects susceptibility is much more difficult to validate. Attempts to investigate any effects on susceptibility may be approached in two ways: either

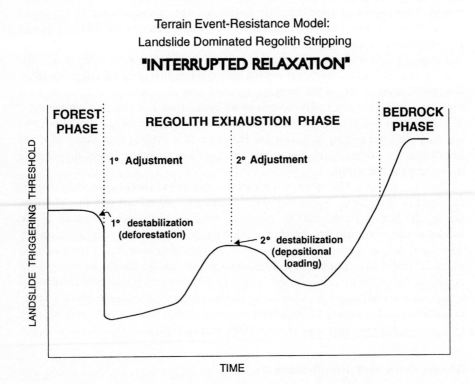

Fig. 2. A conceptual model for New Zealand hill country to account for changes in terrain resistance to landsliding through time. Following deforestation, there is a primary adjustment in terrain resistance indicated by a significant decrease in the landslide–triggering threshold. This is followed by an increase toward a new equilibrium level as a result of regolith stripping. This relaxation toward a new equilibrium level is interrupted, however, by a secondary adjustment within the regolith exhaustion phase. The cumulative effects of depositional loading and/or scarp development once more decrease total catchment stability, resulting in lower landslide–triggering rainfall thresholds which exist until regolith stripping further diminishes available regolith and bedrock begins to dominate resistance.

directly, by examining conditions on–site or, indirectly, by inferring changes from changes in the sediment yield/rainfall threshold, measured off–site. The indirect approach needs to ensure that, in interpreting changes in the rainfall/sediment yield relationship, the effects of antecedent climatic conditions are taken into account before inferences can be made on slope susceptibility. Similarly, complex response and its effect on sediment delivery ratios for different storm characteristics can obscure the actual changes to susceptibility occurring on the slopes.

From geomechanics theory, combined with observation and measurement of landform response to landsliding, a number of mechanisms appear to be operating which will tend to affect both sediment yield from landslide events and inherent susceptibility of the slopes. These are event–induced mechanisms which are an intrinsic effect of the process itself. These were first identified by Crozier (1996) and are further developed below with respect to the proposed model.

Of particular importance to assessing terrain response off–site are the residual conditions of sediment availability and mobility, produced by the landslide event. Changes from event to event in availability and mobility of sediment stored in the area between the failure sites and the sediment yield measurement site can influence the sediment yield for a given magnitude of rainstorm. For example, climatic conditions prevailing during landslide events, on certain occasions, appear to promote full evacuation of the surface of rupture and transport of most displaced material into the drainage networks (Fig. 3). On other occasions, the storm event ends before full evacuation of the landslide site and extensive translocation of material occurs. Thus measured differences in sediment yield attributable to these mechanisms represent a complex response. They indicate variations in catchment delivery capabilities but do not reflect changes in susceptibility to landsliding on the slope.

The mechanisms directly affecting susceptibility on the slope can be placed in two groups:

1. Negative feedback mechanisms which tend to decrease terrain susceptibility.
2. Positive feedback mechanisms which tend to enhance terrain susceptibility.

Negative Feedback Mechanisms

Primary regolith stripping. This involves the reduction of regolith availability (exhaustion model). In the first events after deforestation, the regolith existing in the most susceptible locations is the first to be removed by landslides and consequently the affected terrain retains regolith in the relatively stronger positions. Thus, this mechanism tends to promote a residual increase in inherent resistance ('event resistance'; Crozier 1989) over that prevailing before the event. A similar degree of landsliding in subsequent events may then require an increase in the magnitude of the triggering events, as the catchment resistance and consequently the triggering threshold continue to increase, providing an overall decrease in susceptibility. The concept of event resistance produced by the successive removal of available material has recently been incorporated into what Cruden and Hu (1993) term an

Fig. 3. Shallow regolith landsliding with deposits transported to the stream channel, Wairarapa, North Island, New Zealand.

'exhaustion' model for the prediction of bedding plane rock slides. Dating of landslide scars (Trustrum et al. 1984) indicates that stripping of primary undisturbed regolith proceeds episodically in an upslope direction. The contemporary upslope limit of the regolith stripping process is usually marked by a more or less distinct scarp, formed by the coalescence of contiguous landslide crowns, referred to as the erosion front. The degree of regolith stripping can be directly measured by the percent of the catchment retaining undisturbed regolith and thus can also provide a relative measure of event resistance (Fig. 4). The percentage (relative availability) of undisturbed regolith, as a catchment property, has been used successfully to explain variation in the incidence of landslide between catchments (Crozier et al. 1980).

Depositional hardening. This mechanism is not fully understood, but appears to result from the increase in bulk density that occurs as a result of the remoulding of material as it flows downslope. The increase in bulk density is associated with a measurable increase in apparent cohesion. As this increase in cohesion has been determined from in situ field shear box tests and by back analysis of individual landslides, it may reflect more effective grass root binding in the denser deposits compared with the more porous undisturbed regolith.

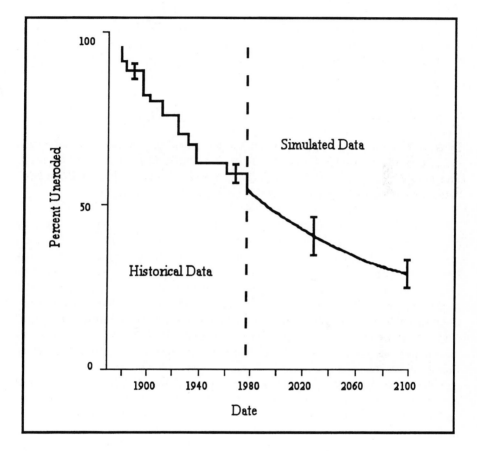

Fig. 4. Extent of regolith stripping by shallow landslides is indicated by the amount of uneroded land measured from sequential photographs up until 1980, and simulated for 100 future events with an annual probability of 0.24 (bars represent standard deviations) (Thomas and Trustrum 1984).

Positive Feedback Mechanisms

Depositional loading. The accumulation of landslide debris, allowing the attainment of critical depth on previously stable sites (Fig. 5).

Scarp development and reduction in toe support. The formation and development of crown and lateral scarps by landsliding has the effect of removing support from the toe of upslope regolith. This factor appears to be influential in the location of subsequent landslides, but its overall affect on terrain susceptibility through time is not known. However, the length of scarp per unit area in a catchment is likely to increase in early episodes of the regolith stripping process then stabilize as an erosion front is formed.

Testing of the Model: Preliminary Results

In an attempt to investigate the extent to which the proposed mechanisms operate to affect susceptibility, two field areas have been studied: Makahu, in Taranaki, in the west of the North Island, New Zealand, and Waikopiro, a sub–catchment of Lake Tutira, east coast, North Island, New Zealand.

The approach taken at Makahu was developed from work (Crozier et al. 1990) which had demonstrated that achievement of critical depth is a necessary precondition for many rainstorm–triggered regolith failures. The aim, therefore, was to investigate the extent to which the landsliding process affected the distribution of depth with respect to 'critical depth' and thus assess susceptibility to future landsliding. Regular sampling was carried out along selected slope profiles to establish regolith depth/slope angle conditions for both undisturbed primary regolith and landslide colluvium. These were then assessed against the critical slope/depth conditions for the respective materials (Figs. 6(a,b)), using strength parameters (cohesion, and angle of shearing resistance) derived from field shear box tests.

The results indicate that the landslide process has dramatically decreased susceptibility. The main mechanisms operating appear to be regolith thinning and increase in both bulk density and cohesion as a result of remoulding of material during transport. Decreased susceptibility is indicated in Figure 6(a,b) by the number of sites that would become unstable if a perched water table reached the surface. Empirical evidence, however, suggests that landslides generally occur when the water table is about one quarter the depth of the regolith (Fig. 7). With respect to the proposed model, the Makahu catchment appears to be in the primary adjustment phase. Either colluvial deposition areas have not yet received sufficient debris from upslope to have regained critical depth or the configuration of the catchment leads to extensive export of debris from the system.

In the Waikopiro catchment, the methodology involves using a GIS to calculate factors–of–safety for each of 26,589 5m x 5m cells in the entire catchment (55 ha in area) as they existed before a major landslide event (Cyclone Bola, March 1988), and comparing these with the distribution of factors–of–safety associated

Fig. 5. A representative colluvial regolith profile. Note loading onto buried soil.

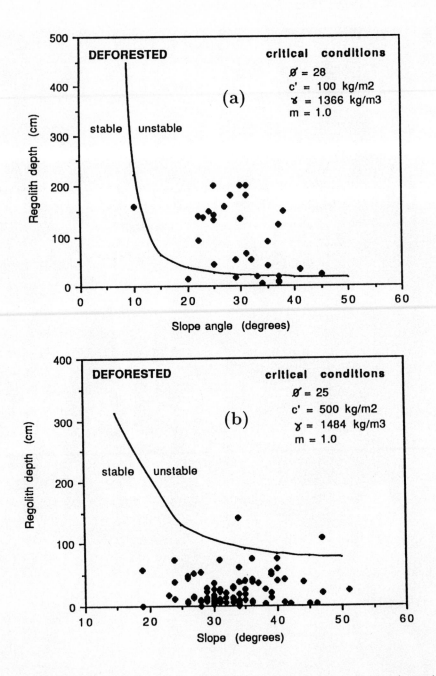

Fig. 6. Comparison of slope angle/regolith depth conditions at sample sites (points), for (a) undisturbed regolith and (b) mass movement colluvium, Makahu, North Island, New Zealand. Critical conditions for landsliding are indicated (line) for listed properties: ϕ, angle of shearing resistance; c, cohesion; γ, bulk density; and an assumption for the ratio of water table depth to depth of regolith (m).

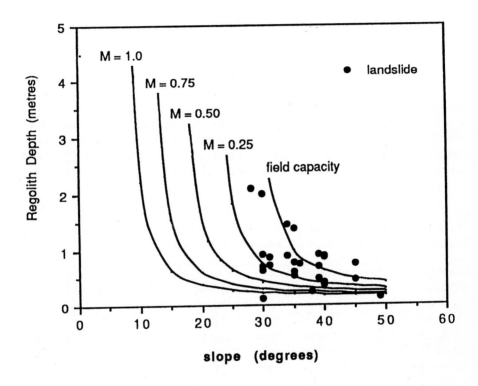

Fig. 7. Comparison of slope angle/regolith depth conditions for actual landslides (points) in previously undisturbed regolith, with critical conditions (lines) for different assumptions of water table height. Soil properties are the same as in Fig. 6(a). Makahu, North Island, New Zealand. Data suggest that water tables of approximately one quarter of regolith depth are critical for landslide occurrence.

with the residual condition, after the event. To achieve this it was necessary to establish a geomechanically relevant classification of the different land surface conditions that exist in a landslide affected terrain. Landsurface classes were then characterized by their geomechanical properties by field and laboratory measurement. Mapped changes in the distribution of these land classes resulting from the event enabled the recalculation of the consequent factors–of–safety. For a full account of the methodology involved see Preston (1996).

Preliminary analysis of the results show that there are significant differences in geomechanical properties between certain of the landsurface classes (Table 1). Changes in the distribution of these landsurface classes could therefore be expected to produce changes in susceptibility, as measured by the factor–of–safety. Table 1 shows that existing landslide scars and new scar surfaces have only a very thin cover of regolith, consistent with original evacuation to bedrock surface. In contrast to the Makahu catchment, depositional loading is evidenced by the increase in thickness of the three landsurface types where recognizable landslide debris has accumulated ('debris on undisturbed', 'debris on colluvium', and 'debris on old scars'), compared with their debris free equivalents. In addition, depositional hardening is indicated in Table 1, by increase in bulk density and cohesion at most sites compared with the source material of undisturbed regolith.

Table 1. Selected geomechanical properties used to characterize landsurface condition classes for terrain units within the Waikopiro catchment.

Landsurface Condition Class	Regolith Depth (m)	Dry Density (kg/m^3)	Saturated Density (kg/m^3)	Apparent Cohesion (kg/m^2)	Angle of Internal Friction (tanform)
Undisturbed	0.65	1180.2	1673.6	215.63	$(\sigma^{-0.7724})*73.097$
New Scar	0.09	1404.4	1821.4	239.06	$(\sigma^{-0.8708})*131.95$
Debris (Undisturbed)	1.18	1257.4	1722.3	215.63	$(\sigma^{-0.7724})*73.097$
Debris (Colluvium)	1.44	1301.0	1753.9	277.53	$(\sigma^{-0.8196})*96.412$
Debris (Old Scar)	0.84	1322.8	1766.0	278.70	$(\sigma^{-0.4681})*8.7992$
Old Scar	0.31	1270.8	1737.2	278.70	$(\sigma^{-0.4681})*8.7992$
Colluvium	0.91	1270.8	1737.2	277.53	$(\sigma^{-0.8196})*96.412$
Alluvium	3.00	1333.2	1782.6	3.64	$(\sigma^{-0.0368})*0.5313$

σ represents normal stress, measured in units of kg/m^2.

Changes in the distribution of landsurface condition between 1965 and 1988 are clearly evident in Table 2. Only 41% of the original forest (primary) regolith remained undisturbed in 1965 and further landslide events, including the 1988 Cyclone Bola event, had reduced this to 33% by 1988. Such results not only indicate the rate of regolith stripping but also provide some resolution of the time scale for the proposed model. In 1988, the remaining 67% of the landsurface is occupied by either erosional forms or depositional surfaces –15% as either recognizable scars or landslide debris and 50% as colluvial surface. It is uncertain as to how much of the colluvial surface can be attributed to deposition from landslides occurring during the post–European settlement phase but recent landslide activity indicates that it is still increasing as a proportion of the catchment area. Overall the results suggest that the Waikopiro catchment is in a much more advanced stage of regolith stripping than the Makahu catchment. In particular depositional loading appears to be an active mechanism in influencing susceptibility.

Table 2. Comparative distributions of landsurface condition classes, 1965 and 1988. Waikopiro, North Island, New Zealand.

Landsurface Condition	1965		1988		% Change
Class	No. of Cells	%	No. of Cells	%	
Undisturbed	10847	40.79	8798	33. 09	−18.9
New Scar	641	2.41	671	2.52	+4.7
Debris (Undisturbed)	127	0.48	78	0.29	−38.6
Debris (Colluvium)	1099	4.13	1244	4.68	+13.2
Debris (Old Scar)	4	0.02	18	0.07	+350.0
Old Scar	1757	6.61	2062	7.76	+17.4
Colluvium	11600	43.63	13287	49.97	+14.5
Alluvium	514	1.93	429	1.61	−16.5
	26589	100.00	26589*	100.00	

* This total includes two cells representing surface water.

It is evident that some of the mechanisms operating will increase the susceptibility of certain sites while decreasing susceptibility of others. The regolith stripping model presented essentially integrates conditions for the whole catchment. One way then of testing the model and determining where the catchment is within the relaxation progression, is to analyse changes in the overall distribution of factors–of–safety on a catchment–wide basis (Fig. 8). The factors–of–safety shown in Figure 8 have been calculated for each of the 26,589 grid cells using the infinite limiting equilibrium method of Henkel and Skempton (1954) and assuming a catchment wide water table depth to regolith depth ratio of m = 0.75. The factor–of–safety is not used here as an absolute value of site stability, because only average values for parameters have been employed. However, it is considered to be a consistent and reliable indicator of changes in inherent resistance.

For both the 1965 and the 1988 condition, by far the majority of the catchment has factors–of–safety below 3.0, with a tail of high values representing evacuated slip surfaces, alluvial flats, and summits. Except for very low values, 1988 has seen a shift in the distribution of factors–of–safety from the < 3.0 to the > 3.0 classes, indicating an overall increase in inherent resistance. However, in Figure 8(b) there is an indication that by 1988 there had been a small increase

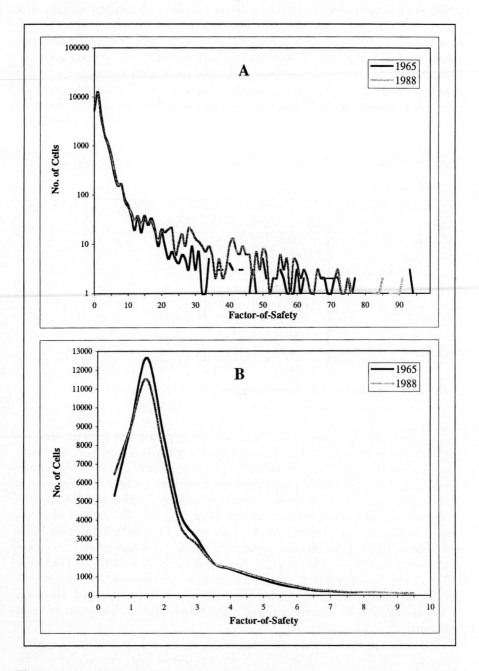

Fig. 8. Frequency distributions of factors–of–safety, 1965 and 1988. (a) All cells (b) Cells with factors-of-safety lower than 10.0 (98.3% and 97.4% of cells for 1965 and 1988 respectively). Waikopiro, North Island, New Zealand.

in the number of cells with very low factors–of–safety. Therefore, it is possible that a small magnitude triggering event, affecting only the most unstable sites, would produce a greater response after 1988 than would have been the case in 1965, indicating increased susceptibility to small storms. However, for high magnitude storms which could destabilize sites with factors–of–safety greater than 1.0, the response today is likely to be significantly less than in 1965, suggesting an increase in inherent resistance with time.

For the whole of the Tutira catchment (within which Waikopiro is located) off–site measurements of sediment yield from two exceptionally large storms suggest that increased inherent resistance (decreased susceptibility) may have been a major mechanism operating between A.D. 1938 and A.D 1988 (Page at al, 1994). Cyclone Bola, in 1988, produced the largest rainstorm on record (753mm over four days); despite its size it produced only about half as much landslide sediment as the smaller ANZAC Day storm of 1938 (692mm over four days). However, as previously discussed, off–site measurements are subject to complex response and are therefore less reliable indicators of terrain susceptibility than on–site assessment.

Conclusion

In unstable terrain subject to episodic landslide activity, assessments of historical or future rates of landslide activity based on empirically established landslide triggering thresholds must be used with caution. There is sufficient evidence to demonstrate that susceptibility to landsliding is affected by event–related changes in terrain resistance. In the early stages of regolith removal from the landscape, regolith stripping to bedrock and depositional hardening of deposits increase the triggering threshold. However, as the erosional process continues over a period of decades, accumulation of deposits on lower slopes tends to lower temporarily terrain resistance, before subsequent thinning of the regolith causes an ultimate increase in terrain resistance.

Acknowledgment. We wish to thank the New Zealand/ Federal Republic of Germany Scientific and Technological Cooperation Agreement for a grant in aid to support this work.

References

Crozier, M.J. (1989): Landslides: Causes, Consequences and Environment. Routledge.
Crozier, M.J. (1996): Magnitude/frequency issues in landslide hazard assessment. In: R. Mäusbacher and A. Schulte (Eds.) Beitrage zur Physiogeographie. Barsch Festschrift, Heidelberger Geographische Arbeiten 104:221–236.
Crozier, M.J., R.J. Eyles, S.L. Marx, J.A. McConchie, and R.C. Owen (1980): Distribution of landslips in Wairarapa hill country. New Zealand Journal of Geology and Geophysics 23:575–586.

Crozier, M.J., M. Gage, J.R. Pettinga, M.J. Selby, and R.J. Wasson (1992): Stability of hillslopes. In: J. Soons and M.J. Selby (Eds.) Landforms of New Zealand. Second Edition, Longman, 63–90.

Crozier, M.J., and B.J. Pillans (1991): Geomorphic events and landform response in south–eastern Taranaki. Catena 18(5):471–487.

Crozier, M.J., E.E. Vaughan, and J.M. Tippett (1990): Relative instability of colluvium–filled bedrock depressions. Earth Surface Processes and Landforms 15:326–339.

Cruden, D.M., and X.Q. Hu (1993): Exhaustion and steady state models for predicting landslide hazards in the Canadian Rocky Mountains. Geomorphology 8:279–285.

Glade, T., and M.J. Crozier (1996): Towards a national landslide information base for New Zealand. New Zealand Geographer 52(1):29–40.

Glade, T., and M.J. Crozier (in press): Rainfall–triggered landslides in New Zealand: Bibliography. Journal of Environmental Geology.

Greenway, D.R. (1987): Vegetation and slope stability. In: M.G. Anderson and K.S. Richards (Eds.) Slope Stability, Wiley, 187-230.

Henkel, D.J., and A.W. Skempton (1954): A landslide at Jackfield, Shropshire, in an overconsolidated clay. Proc. Conf. Stability of Earth Slopes, Stockholm 1, 90–101.

O'Loughlin, C.L., and A.J. Pearce (1976): Influence of Cenozoic geology on mass movement and sediment yield response to forest removal, north Westland, New Zealand. Bulletin of the International Association of Engineering Geologists 14:41–46.

Page, M.J., and N.A. Trustrum (1997): A late Holocene lake sediment record of the erosion response to land use change in a steepland catchment, New Zealand. Zeitschrift für Geomorphologie 41(3):369-392.

Page, M.J., N.A. Trustrum, and R.C. DeRose (1994): A high resolution record of storm-induced erosion from lake sediments, New Zealand. Journal of Paleolimnology 11:333–348.

Preston, N.J. (1996): Spatial and Temporal Changes in Terrain Resistance to Shallow Translational Regolith Landsliding. Unpublished MSc.(Hons) thesis in Physical Geography, Victoria University of Wellington.

Selby, M.J. (1976): Slope erosion due to extreme rainfall: a case study from New Zealand. Geografiska Annaler 58(A):131–138.

Selby, M.J. (1993): Hillslope Materials and Processes. Second Edition. Oxford University Press.

Thomas, V.J., and N.A. Trustrum (1984): A simulation model of soil slip erosion. Sym. Effects of Forest Land Use on Erosion and Slope Stability. Environment and Policy Institute, East–West Center, University of Hawaii, Honolulu, 83–89.

Trustrum, N.A., V.J. Thomas, and G.B. Douglas (1984): The impact of forest removal and subsequent mass-wasting on hill country pasture productivity. Sym. Effects of Forest Land Use on Erosion and Slope Stability. Environment and Policy Institute, East-West Center, University of Hawaii, Honolulu, 308–309.

Wilmshurst, J.M. (1997): The impact of human settlement on vegetation and soil stability in Hawke's Bay, New Zealand. New Zealand Journal of Botany 35(1).

Precision of Parameter Estimation in Meander Models

C. Droste

Institute of Photogrammetry, University of Bonn, Germany

Abstract. This paper is organised in two sections. In the first section a dynamical model for meander migration is developed. It includes one dimensional fluid dynamics (main velocity and secondary flow) and a simple approach for migration mechanisms. The second section is concerned with generic aspects of parameter estimation in nonlinear dynamical systems. It shall be shown that the experimental design is essential for the precision of estimated parameters. Especially the choice of initial conditions and observation times is examined for the example model of meander migration. The methods can be used to improve the precision of calibration for general nonlinear dynamical models.

1 Introduction

The phenomenon of meander generation and its evolution has been subject to a large number of studies. A common approach is the establishment of some phenomenological relations in the form of power laws (e.g. Leopold and Maddock 1953, Ackers and Charlton 1970, Schumm 1963) based on field studies of some rivers. This general description was seldom satisfactory and deviations had to be explained by the local environment (Ferguson 1975).

The desire to find fundamental and more physically based reasons for meandering paved the way for very different, but incompatible theories. Yang (1971) tried to explain meandering by minimal energy expenditure per water mass and Leopold and Langbein (1966a, 1966b) regarded a river as random walk, finding its path by minimising the sum of the squares of the changes in direction. Such variational principles seem to be more fundamental, but their compatibility with the basic laws of mechanics is not proved. Thus, in spite of their possibly correct description of some meander characteristics, they remain as phenomenological as power laws do.

Only the work of Parker (1976) led to a more physical understanding of meandering and resulted in the fundamental works of Ikeda et al. (1981) and Parker et al. (1982). On the background of these works genuine dynamical models were developed by Johannesson and Parker (1989) and Howard (1992, 1995). Even the effects of the secondary flow have been included. Its importance had been pointed out in theoretical and experimental works (e.g. Bernard and Schneider 1992, Zimmermann 1974). General testing of these dynamical models over a range of hydrologic and geologic conditions has been performed by Cherry et al. (1996).

In spite of their relative success, such models remain artificial in that they claim a dependence of the local erosion rate from a near bank perturbation velocity or perturbation depth, i.e., the difference between those quantities near the bank and the average values, but independence of the average quantities themselves. The deficiencies of those models have been pointed out in the work of Cherry et al. (1996).

Comparison of model predictions and natural evolutions still lacks in various points:

- Obviously, the models cannot contain all influences of the environment. Thus, additional (random) processes have to be claimed for the compatibility of predictions and actual evolutions. Therefore dynamical models are more likely to describe laboratory experiments, where many of these unmodeled influences can be avoided (precipitation, vegetation, human impact etc.). They may be more suited to examine the basic processes.
- Model specific parameters have to be fitted, to admit proper predictions (calibration). This is seldom done with statistically appropriate data sets.
- The sensitivities of the outcomes to the parameters and the initial conditions must be taken into consideration to assess the differences between observation and prediction.

The present work develops a model that is complex enough to generate a realistic behavior and focuses on the last two points, searching adequate conditions for an appropriate parameter estimation. The presented model was aimed to meet both, a consequent modeling of qualitatively observed phenomena and involved processes and a reliable physical background. A reasonable simplicity of the model had to be kept to allow the performance of subsequent examinations on the sensitivity to parameter changes and initial conditions.

2   Meander Model

For the development of a new meander model it was considered essential to include the secondary (helical) flow and use fundamental fluid mechanical laws for the description of the fluid dynamics. This results from the qualitative description of geomorphological works (e.g. Markham and Thorne 1992).

A draft of the processes taking place may be considered as follows. The water flowing in an existing channel adjusts its surface level and velocity according to the given geometrical conditions, i.e., the slope of the flood plane, the actual direction of the channel, the discharge and the friction forces. The asymmetry of the cross sections found in a bend results from asymmetric fluid dynamics, viz. the secondary flow. Joined with the main flow it builds a helical flow pattern. It is induced by the different centrifugal forces at the water surface and near the bed producing a tortuous net force.

The secondary flow causes different processes at both banks: erosion on the outer cut–bank because it acts in addition to gravity, and deposition of the material at the inner point bar because it counteracts gravity. The sign of the helical motion changes due to the different signs of curvature of the thalweg. However, a sinusoidal river will have a helical flow pattern, moderately shifted to the phase of the meander because the water's inertial forces prevent the system from instantaneous adaption to the geometric conditions. This causes the observed inherent asymmetry of meander planforms (Carson and Lapointe 1983).

The model contains three parts, viz. geometry, fluid dynamics and sediment transport. The geometry of the river is given by a polygon (enbeded in an inclined plane) describing the thalweg of the meander. At each point of the polygon the actual cross section is represented by two quadratic functions, building the left and the right bank. Depending on which side the actual cut–bank is, one of the functions has the higher coefficient resulting in a steeper slope (Fig. 1). Thus, together with the depth H of the channel, three parameters govern the shape of the cross section. The *fluid dynamical* description aims at differential equations

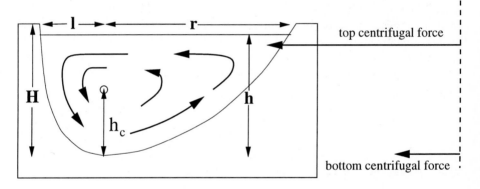

Fig. 1. Cross section with secondary flow and different bank slopes. The three parameters l, r and H determine its shape.

for the average main velocity $\widetilde{u}$, the height of the water level h, and the average secondary flow v_c. The problem requires a special form of the *continuity equation*, which shall be treated for incompressible flow ($\frac{\partial \rho}{\partial t} = 0$):

$$\nabla \cdot (\rho \boldsymbol{u}) = \frac{\partial \rho u}{\partial x} + \frac{\partial \rho v}{\partial y} + \frac{\partial \rho w}{\partial z} = 0 \ . \tag{1}$$

Herein x, y, z is a local orthogonal coordinate system with x pointing into the direction of the thalweg, y pointing perpendicularly into the inclined plane and u, v, w are the velocities in x–, y– and z–direction, respectively. The continuity equation for the one–dimensional problem of a river may be denoted as

$$\frac{\partial}{\partial x}(\rho A \widetilde{u}) = 0 \quad \Leftrightarrow \quad \widetilde{u}\,\frac{\partial A}{\partial x} + A\frac{\partial \widetilde{u}}{\partial x} = 0 \ , \tag{2}$$

using the notation $\tilde{\bar{u}}$ for the velocity u that is averaged horizontally ($\tilde{}$) and vertically ($\bar{}$) across the river. A is the cross section containing the fluid.

For a description of the *main flow* in the direction of the thalweg, we inspect the cross section integrated Navier–Stokes equation in x–direction

$$\underbrace{\int_{y_L}^{y_R}\int_{z_B}^{z_S} \frac{\partial u}{\partial t} + u\frac{\partial u}{\partial x} + v\frac{\partial u}{\partial y} + w\frac{\partial u}{\partial z}\, dzy}_{I_1} = \tag{3}$$

$$= \underbrace{\int_{y_L}^{y_R}\int_{z_B}^{z_S} \frac{-1}{\rho}\frac{\partial p}{\partial x}\, dzy}_{I_2} + \underbrace{\int_{y_L}^{y_R}\int_{z_B}^{z_S} \frac{\partial}{\partial x}\nu\frac{\partial u}{\partial x} + \frac{\partial}{\partial y}\nu\frac{\partial u}{\partial y} + \frac{\partial}{\partial z}\nu\frac{\partial u}{\partial z}\, dzy}_{I_3} \;,$$

where z_B, z_S, y_L and y_R are the z–coordinates of the bottom and surface and the y–coordinates of the left and right bank, respectively. The order of integration is of importance because in a non rectangular bed the z–coordinates depend on the y–coordinate. The kinematic viscosity ν is defined as the quotient of the dynamical viscosity and the density η/ρ. The integration can be performed term–wise (cf. Malcherek 1995). With some assumptions the left side of Eq. (3) turns into

$$I_1 = A\left(\frac{\partial \tilde{\bar{u}}}{\partial t} + \tilde{\bar{u}}\frac{\partial \tilde{\bar{u}}}{\partial x}\right)\;. \tag{4}$$

The pressure term in Eq. (3) can be evaluated to

$$I_2 = \int_{y_L}^{y_R}\int_{z_B}^{z_S} \frac{-1}{\rho}\frac{\partial p}{\partial x}\, dzy = -\int_{y_L}^{y_R} gh\left(\frac{\partial z_B}{\partial x} + \frac{\partial h}{\partial x}\right)\, dy = -gA\left(\frac{\partial \tilde{z_B}}{\partial x} + \frac{\partial \tilde{h}}{\partial x}\right) \tag{5}$$

if the hydrostatic pressure approximation is used. The difference between water surface and the bottom is labeled $h = z_S - z_B$.

The viscid terms (I_3) of Eq. (3) are treated similarly to I_1 and yield

$$I_3 = \int_{y_L}^{y_R} \nu\frac{\partial u}{\partial x}\bigg|_{z_B}\frac{\partial z_B}{\partial x} + \nu\frac{\partial u}{\partial y}\bigg|_{z_B}\frac{\partial z_B}{\partial y} - \nu\frac{\partial u}{\partial z}\bigg|_{z_B}\, dy\;, \tag{6}$$

if laminar friction and turbulent friction at the surface (to the air) is negligible. The integrand can be considered as a scalar product of the normal vector at the bottom with the specific stress tensor because only the u–component of the velocity is considered. The gradient ∇u is also normal to the bottom because u vanishes there and hence tangential components have to be zero. Thus, the stress vector can be written as a product of the shear stress $\tau_x(u)$ multiplied by the normal unit vector n, so that the integral appears to be

$$\int_{y_L}^{y_R} -\nu \left.\frac{\partial u}{\partial x}\right|_{z_B} \frac{\partial z_B}{\partial x} - \nu \left.\frac{\partial u}{\partial y}\right|_{z_B} \frac{\partial z_B}{\partial y} + \nu \left.\frac{\partial u}{\partial z}\right|_{z_B} dy \tag{7}$$

$$= \int_{y_L}^{y_R} \begin{pmatrix} \frac{\partial z_B}{\partial x} \\ \frac{\partial z_B}{\partial y} \\ -1 \end{pmatrix} \cdot \tau_x(u) \frac{1}{\sqrt{\left(\frac{\partial z_B}{\partial x}\right)^2 + \left(\frac{\partial z_B}{\partial y}\right)^2 + 1}} \begin{pmatrix} \frac{\partial z_B}{\partial x} \\ \frac{\partial z_B}{\partial y} \\ -1 \end{pmatrix} dy$$

$$= \int_{y_L}^{y_R} \tau_x(u) \sqrt{\left(\frac{\partial z_B}{\partial x}\right)^2 + \left(\frac{\partial z_B}{\partial y}\right)^2 + 1} \, dy$$

$$\approx \tau_x(u) \int_{y_L}^{y_R} \sqrt{\left(\frac{\partial z_B}{\partial y}\right)^2 + 1} \, dy = \tau_x(u) \int_{y_L}^{y_R} dl_u = \tau_x(u) \, l_u$$

$$= \tau_x(u) \frac{A}{r_{hy}}.$$

Here, the *wetted perimeter* l_u and the *hydraulic radius* $r_{hy} = A/l_u$ have been introduced. The approximation in the second step was possible because the slope in x–direction (≤ 0.01) is small compared to that in y–direction and 1 ($\frac{\partial z_B}{\partial x} \ll \frac{\partial z_B}{\partial y}, 1$).

Collecting the results of Eqs. (4)–(7), the following form of Eq. (3) is obtained.

$$\underbrace{A \left(\frac{\partial \tilde{u}}{\partial t} + \tilde{u} \frac{\partial \tilde{u}}{\partial x}\right)}_{I_1} = \underbrace{-gA \left(\frac{\partial \tilde{z}_B}{\partial x} + \frac{\partial \tilde{h}}{\partial x}\right)}_{I_2} + \underbrace{\frac{\tau_x}{r_{hy}} A}_{I_3} \tag{8}$$

On fluid dynamical time scales the problem shall be treated stationary. Hence we can drop the temporal derivative (as well as the markers for the spatial averages).

Equating the representation of $\frac{\partial A}{\partial x}$ from the continuity equation and application of the chain rule yields

$$\frac{\partial h}{\partial x} = -\frac{A}{bu}\frac{\partial u}{\partial x} - \frac{1}{b}\sum_i \frac{\partial A}{\partial c_i}\frac{\partial c_i}{\partial x} , \tag{9}$$

where b is the width of the channel at the water surface and c_i ($i = 1\ldots3$) are the three parameters describing the cross section geometry (Fig. 1). As a result the right side of Eq. (8) can now be expressed as a function of the velocity and the geometry only, while Eq. (9) will yield the differential equation for the second unknown h.

For the remaining turbulent friction force I_3 we introduce the following formula containing the constant C_1 which may be subject to parameter estimation and substitute the commonly used approach.

$$\tau_x \approx -C_1 \, u|u| \tag{10}$$

C_1 may also be chosen to be dependent on the hydraulic radius ($C_1 = C'_1\, r_{hy}^{\frac{1}{6}}$) to achieve consistency with the Manning formula for channels with constant slope, whilst the given formula corresponds to Chezy's approach (Hsü 1989).

Substitution of Eqs. (9) and (10) into Eq. (8) and rearrangement of terms results in the final differential equation for the averaged velocity u.

$$\frac{\partial u}{\partial x} = \frac{-bu}{bu^2 - gA}\left(g\left(\frac{\partial z_B}{\partial x} - \frac{1}{b}\sum_i \frac{\partial A}{\partial c_i}\frac{\partial c_i}{\partial x} \right) + C_1\frac{u|u|}{r_{hy}} \right) \tag{11}$$

It is substituted into Eq. (9) to obtain the final form of the differential equation for the height of the water level h. This is equivalent to the *Saint–Venant equation* (Malcherek 1995) for channel flows.

Finally the *secondary flow* shall be determined to obtain a quantitative description of the fluid dynamical state, as it was qualitatively described earlier. A rigorous derivation of the secondary flow from the Navier–Stokes equation would require the solution of the entire 3D problem because a separation like that of the main velocity cannot be achieved due to the lack of a proper symmetry and the dependence on the main velocity.

Hence the separation is achieved by a hierarchical view on the problem. The velocities perpendicular to the main flow are one to two orders of magnitude smaller than the main flow itself. Thus, the influence of the secondary flow on the averaged main flow can be considered negligible, whilst helical flow patterns are induced by the primary flow and the geometry. Consequently, the averaged main velocity can be determined from the integration of Eq. (11) only, and the differential equation for the secondary flow must contain u and additional information about the vertical flow pattern.

It has been pointed out that the motion transversal to main flow is generated by the differences of the centrifugal forces in different depths. Therefore, an approximate distribution of the main velocities has to be known, although it has not been determined by Eq. (11). This distribution is modeled as proposed for a rectangular cross section by Allen (1992):

$$u(y, z) = U_c\left(\frac{z - z_B(y)}{h} \right)^n, \tag{12}$$

where U_c is a constant normalising factor and n is a small exponent comparable with $\frac{1}{7}$. This distribution forces the maximal primary velocity to be lying at the water surface above the deepest cross section position.

The circular flow itself has an unknown pattern as well, but it should be similar to that of Fig. (1), independent of the magnitude of its velocity. Hence the driving force for a volume element is the projection of the centrifugal force on the local direction of (possible) flow given by this pattern. Allen (1992) used a simple partitioning of the water body into the upper and lower half, distinguishing only driving (factor 1) and resisting (factor -1) forces. A slightly refined model shall be presented here, dividing the cross section into three areas delimited by the straight lines pointing from the center of torsion (lying h_c above the deepest point) to the left bank, the right bank and downwards, respectively. As shown in Fig. (2), the directions of projection are taken to be the direct

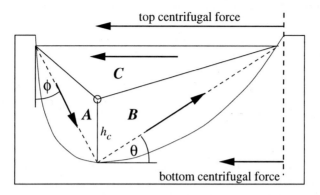

Fig. 2. Cross section divided into three parts (A, B, C) allowing a coarse modeling for the secondary flow.

connections between the extremal points, i.e., the weight factor k for the different contributions to the net centrifugal force is given by

$$k(y, z) = \begin{cases} -\sin\phi & y, z \in A \\ -\cos\theta & \text{for} \quad y, z \in B \\ 1 & y, z \in C \end{cases} \tag{13}$$

To obtain a differential equation for an averaged secondary flow v_c, the centrifugal force must be integrated over the wetted cross section and the turbulent friction force must be subtracted.

$$\frac{\partial v_c}{\partial x} = \frac{1}{A(x)u} \int \int \frac{u^2(y, z)}{R(x)} k(y, z) dA - \frac{C_2}{r_{hy}} v_c^2 \tag{14}$$

$R(x)$ denotes the local radius of curvature of the thalweg and C_2 is another friction coefficient.

For completion of the dynamical meander model the treatment of *sediment transport* is still missing. The so far developed Eqs. (9, 11, 14) are a system of coupled ordinary differential equations and its integration yields the values for u, h, v_c as a function of the thalweg length x if the water supply is given. These values shall now be used to calculate local meander migration.

The possible kinds of river load can be classified into bed load (dragging and saltation), suspended load and dissolved load. Although bed load is very important in natural rivers, only suspended load will be taken into account because we focus on laboratory experiments. The scouring effect of suspended load is similar to that of the fluid itself and can be assumed to be depending only on fluid dynamical properties and geometry. Therefore the sediment concentration in the flow is not modeled explicitly. It is assumed that the river yields a constant sedimentary load from the upper stream areas, that is high enough not to be seriously affected by the local processes. Although this assumption may be wrong in detail, it holds for longer time scales because meander migration is dominating processes narrowing and widening the channel, i.e. a certain segment of the river will approximately maintain its cross section area during migration being neither sink nor source for sediment.

Detailed modeling of erosion and deposition is in general an unsolved problem. The crucial quantity is the shear stress τ, partly arising from the velocity of the adjacent water and partly from gravity acting on the sediment in steep locations. Although there exist some empirical approaches that model erosion and deposition depending on the local conditions, a practical point of view shall be taken. Anyway, there is no generic formula that determines shear stresses under generic circumstances from fluid dynamical properties. Since we are not interested in detailed changes of the form of the cross section we may assume that migrations of the thalweg are directly influenced by fluid dynamics. The approach assumes that it is proportional to the following combination of the main velocity, the circular velocity and the curvature of its path.

$$q = C_3 u^2 + C_4 v_c^2 + C_5 \frac{1}{R} \tag{15}$$

Here, q is the migration velocity of the thalweg perpendicular to the tangential direction. Although this relation has no hydrologic foundation, generated evolutions in computer experiments look quite natural.

We have presented a model for predicting the temporal migration of meander thalweg lines. For the following examinations it is not important whether the model is able to describe real channels or not. We only want to know how measurements should be performed to guarantee the highest possible precision of parameter estimation *if* the mathematical model is applicable to a natural system. In other words, the measurements should extract as much information as possible *within* this model. The methods presented in Section 4 are applicable whether the model represents real or hypothetical evolutions. Before the question can be answered, the meander model has to be regarded in the context of nonlinear dynamical systems.

3   The Meander Model as a Nonlinear System

The meander model shall now be considered as a *nonlinear dynamical system*, which is generally represented by a differential equation of the type

$$\dot{x} = f(x, \alpha) \ . \tag{16}$$

The state of the system is described by a vector x; the number of its components is given by the number of the model's degrees of freedom. It changes in time according to a (nonlinear) function of itself and a parameter vector α. Regarding the geometry of the meandering river as the state of the system, this pattern is fulfilled because the future evolution of the geometry is governed by its present state and some constant parameters. The parameters $C_1, \ldots, C_5$ from the previous section build the parameter vector $\boldsymbol{\alpha}$. Although there are other important variables, like the discharge and the sediment yield (influenced by precipitation and the erosivity of the upper stream parts), these shall be treated as constants because they are controllable in laboratory flume studies.

The temporal derivative of the geometry has not been developed as an analytical function as in (16). However, it is possible to carry out numerical computer experiments implementing the equations of the meander model. Thus it is possible to realize a numerical approximation of a relation equivalent to Eq. (16). However, some simplifications are necessary to make the numerical representation of system states treatable.

As it was pointed out earlier, the essential fluid dynamical differences on the outer and the inner bank of the river are induced by centrifugal forces, which emerge from the curvature of the thalweg line and cause secondary flows. The relaxation of some arbitrary cross section shape to the (natural) shape induced by the curve pattern of the center line through the helical flow is only a matter of relative short time scales, and the feedback of small changes in the cross section to the center line is limited. Hence the cross section is a subsidiary component of the system and may be neglected resulting in a coarser description of the system. Therefore the representation of the states by a curved line and the empirical neglect of the detailed erosion and deposition laws is appropriate.

For further reduction of the data describing a system's state, the line is represented by the local direction $\phi(x)$ of the river's thalweg in the plane as a function of the length x of the path. Given a sine–generated curve, i.e., $\phi(x) = A \sin(\omega x)$, with reasonable values for ω and A as an initial state, the absolute values of the complex Fourier spectrum are all zero except the one representing the sinuosity with the spatial frequency ω. Such a system typically enlarges its ω–component and generates new components. The smoothness of $\phi(x)$ may now be exploited by building the Fourier transform. High frequency components are of low interest. They are only affected by random processes. Instead, the interesting behavior is observed in the components with lower frequencies, containing the dynamic of the deterministic model given above. Several examinations showed that about 15–30 Fourier coefficients were sufficient to represent the behavior of the channels under consideration.

We are seeking initial conditions that grant optimal parameter estimation from measured evolutions. This implies low sensitivity of the outcomes to the initial conditions and high sensitivity to parameters that shall be estimated. In the next section two initial conditions shall be compared, one with $A \approx 0$ (i.e. a channel initially prepared straight) and one with $A = 0.3, \omega = 0.5$ (i.e. a slightly curved "snake"). The examined curves represent channels that are a few meters long. We want to examine which initial condition is better suited for the estimation of C_4, while the other constants are fixed at $C_1 = 0.2$, $C_2 = 2.5$, $C_3 = 0$ and $C_5 = -10^{-4}$. Since only computer experiments have been performed, the mathematical quantities are presented without units. They stem from visually plausible evolutions.

4   Precision of Parameter Estimation

Generally the determination of parameters in a nonlinear system is performed in two steps. Firstly the system is prepared or (for natural systems) measured in its initial condition. After a certain time t the state of the system is measured again to locate its position in state space. If enough *measurements* have been taken, the values of open parameters can be chosen so that a given criterion (e.g. the least square criterion) is fulfilled. In the case of meander migration the initial condition is the geometry of a channel either prepared in a laboratory experiment or detected by airborne photogrammetric devices. Also the states at the time t are detectable by photogrammetric means in either case.

The *measurements* are principally perturbed by errors. On one hand there are deficiencies of the measurement methods. Therefore the preparation of the system in the initial state as well as the measurement of the outcome after time t are uncertain. On the other hand additional, unmodeled processes called *dynamical noise* act on the system, corrupting the information that shall be drawn from the evolved system.

It depends on the inherent properties of the system whether the errors made during preparation and those stemming from dynamical noise tend to be enlarged or to be suppressed. In high dimensional systems even both may occur, i.e. errors in some directions of the phase space are enlarged and errors perpendicular to those tangential subspaces are decreased during evolution. In any case the resulting errors are added to those which are made when the state of the system is detected at the end of the evolution.

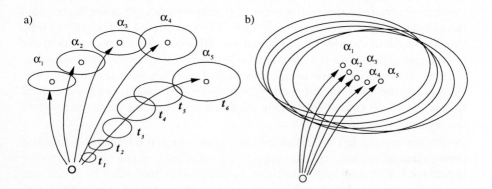

Fig. 3. Schematic evolution of errors compared to parameter sensitivity

Fig. 3 grants an abstract view on the situation. The phase space of the system is given by the plane of the paper. Starting at a given point, measured with finite accuracy, the system evolves according to the values of the parameters. Different parameters are represented by different arrows. For the sake of simplicity think of only one parameter equidistantly sampled in an interval, such that the parameter values α_i are given by $\alpha_i = \alpha_0 + i\,\Delta\alpha$, $i = 1, \ldots, 5$.

The initial system may deviate from the measured initial position because the preparation is inaccurate. The way the distribution of potential initial conditions is propagated depends on the contraction and expansion properties of the system along the individual paths. The ellipses represent *confidence limits*, i.e. the area with a certain (e.g. 99 percent) chance that the outcome of an experiment falls within this region. The form and magnitude of the resulting distributions of measured end states must be compared to the differences between exact evolutions of differently parameterised systems. Since the ellipses do not overlap, it will be most likely to distinguish the parameters correctly from the results of the measurements of the outcomes.

Fig. 3b may by drawn from an other region of the *same* systems phase space. Assumed that the confidence limits have kept their size, the distances between outcomes for different parameters shrunk. The relative magnitude of the confidence areas and the differences between the outcomes are not appropriate any more for a reliable resolution of a parameter differences $\Delta\alpha$.

A method for quantitative assessment of such measurement situations has been developed. It uses stochastic simulations of the evolution to generate an ensemble of possible outcomes after time t. Each of the evolutions is therefore randomly influenced in two ways. On one hand the initial condition is disturbed with Gaussian distributed errors to simulate the limited measurement accuracy. On the other hand random noise is added to the states to simulate dynamical noise.

The state after time t marks the simulated outcome of this probabilistic computer experiment and the complete ensemble represents the distribution of possible states after time t. The ensemble serves to generate a hypothesis for the probability distribution of final states by defining hyper boxes of the state space and counting the frequency of outcomes in each of it. This is an empirical value for the probability of the discretised final states.

These probability distributions are exploited with the help of *information theory* (Shannon and Weaver 1949). Mathematically this is equivalent to the concept of *entropy*. It provides general methods to determine the *information*, contained in a signal stream, if the signals are elements of a finite set of symbols with given probabilities. The symbols correspond to different hyper boxes representing distinguishable outcomes of the experiment. Therefore the size of the boxes must be similar to the measurement errors for detection.

In the context of parameter estimation we are interested in the information about the parameters, resulting from an observed evolution of the system. The concept of *mutual entropy* H (e.g. Vosselmann 1992, Papoulis 1984) delivers such a measure. It determines the information about a random variable $\underline{a}$ with discrete probability distribution $p(a_i)$, $i = 1, \ldots, n$ that is contained in the observation of an other random variable $\underline{b}$ with discrete probability distribution

$p(b_j), \ j = 1, \ldots, m.$

$$H(\underline{a}; \underline{b}) = \sum_{ij} p(a_i, b_j) \ \log \frac{p(a_i, b_j)}{p(a_i)p(b_j)} \tag{17}$$

The two random variables correspond to the parameter vector ($\underline{a}$) and the state vector ($\underline{b}$), respectively. The probability distribution of possible outcomes is generated with random parameters. The parameter distribution reflects the a priori knowledge about the parameters. Consider the only knowledge about a searched parameter is, that it is from the interior of a certain interval. Then the appropriate probability distribution is uniform within this interval and zero outside.

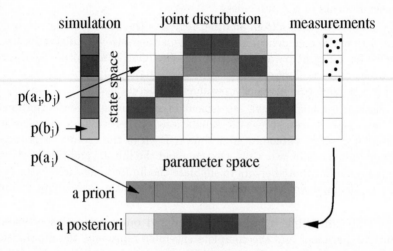

Fig. 4. Example distribution of simulated outcomes in a one dimensional state space (left) partitioned into five boxes, the a priori and a posteriori distributions of the parameter (bottom) and the joint distribution. The darkness indicates the probability of the box (white: $p = 0$, black: $p = 1$).

Fig. 4 shows an example with only one parameter and one dimension in state space. This implies that the hyper boxes in both spaces are intervals. With a growing number of dimension they become rectangles, rectangular boxes, 4D–boxes, etc. Dark intervals correspond to high probabilities. The 'a priori' parameter distribution is uniform indicating the above mentioned knowledge about this parameter.

The stochastic simulation allows to determine an empirical *joint distribution*, i.e. probabilities for each of the discrete parameter values and each of the boxes in state space. The bar on the left symbolises the distribution of the simulated outcomes in 1D cumulated over all parameter values. If measurements are available for equivalent real evolutions (right column in Fig. 4) the knowledge about the parameter can be updated with the *Bayesian rule*. In the example this is indicated by higher 'a posteriori' probabilities for the central intervals. The new distribution corresponds to higher information about the parameter. The difference in knowledge contained in these distributions is the mutual entropy/information H. When the joint distribution is known, H can be computed from Eq. 17.

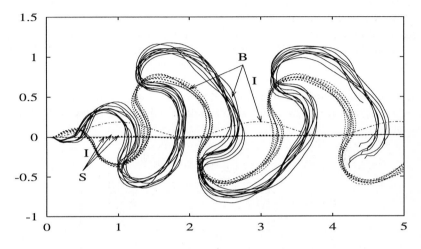

Fig. 5. Comparison of the planform meander evolution for two different parameter values (dashed and solid curves) and different initial states I, one being straight (S) and one bended (B).

The method can be applied to the previously defined meander model. Fig. 5 shows four sets of possible evolutions of a meander center line, viz. all combinations of two parameter values for parameter C_4 and two initial conditions. Each of the sets contains 10 outcomes, selected from an ensemble of 2500 simulations. They give an impression of the variability obtained by the addition of small stochastic processes (to model dynamical noise) and uncertainties in preparing the initial state. The standard deviation of the dynamical noise of 0.015 was applied after each time step.

Two of the sets (S) are almost invisible because the initial state (I) has been chosen to be a straight line and observation time has not been large enough to form river bends. If the simulation is continued to longer periods meander bends appear even for this initial condition because some of the randomly generated deviations from a straight line become large enough to build stable structures. However, in Fig. 5 the observation with limited resolution does not allow a distinction from the initial state. Thus no information about the two differently chosen parameter $C_4 \in \{600, 1000\}$ is won.

The *other two sets* of thalweg lines (B) represent evolutions of an initially (I) bended line, a sine–generated curve with $A = 0.3$ and $\omega = 0.5$. They show much larger variability for each of the two parameter values. However, the sets are clearly separated from each other allowing to distinguish the chosen values for C_4. The parameter determines the coupling between circular velocity and migration velocity. Therefore $C_4 = 1000$ represents a faster evolution than $C_4 = 600$, and the average evolutions for both values diverge. Since the dispersion caused by dynamical noise is lower than that caused by differences in the parameter, a measurement of the final state leads to a clear decision for one of the parameter values.

The heuristic finding must correspond to the temporal evolution of H in both cases. For practical reasons the observation has been restricted to the absolute values of the first seven Fourier components of the meander lines. This was achieved by choosing the box width in all other dimensions large enough to contain all ensemble members. The length of each path had been fixed. For each time step a fixed partitioning of the state space (7 dimensions) was used to calculate the joint distribution and finally the mutual entropy H from 2500 evolutions.

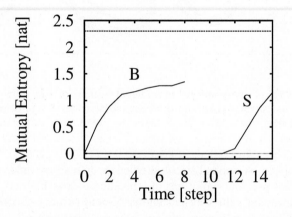

Fig. 6. Mutual Entropy (Information) of the probability distribution of the ensemble and a uniformly distributed parameter in an interval equidistantly partitioned in 10 subintervals.

The results are shown in Fig. 6. The curves indicate that the bended initial condition (B) is superior to the straight (S). For the first 6 time steps the latter yields no information at all and the time spent for such experimental studies does not yield any information about the parameter. The sets of outcomes shown in Fig. 5 are taken from time step 6, where the discrepancy is maximal.

The information gained by observation of the bended river is almost saturated at $t = 6$ (later time steps have been omitted because meander cut off occurs for a considerable part of the ensemble). Even higher values for the mutual entropy could have been achieved with higher accuracies for observation, i.e., finer grained partitioning of the state space, but these results become less reliable, when the ensemble size (2500) is constant. When the state space expands, each particular member of the ensemble will sooner or later end up in its own hyper box, and the ensemble cannot be used to estimate the probability distribution. In the given case the limit entropy for 2500 simulations is marked by the horizontal line at 2.3. Instead, the entropy is only calculated with 10 or more members per hyper box. Lower average numbers are suspicious and should be avoided.

The heuristically found superiority of bended initial conditions over straight ones has been quantitatively verified. There are other possibilities to use the approach of information theory, e.g. the method has been used to determine optimal measurement points for situations where only isolated points can be used to detect meander migration, and the integral view of Fourier coefficients is not appropriate. However, for all applications a good guess about the measurement design that shall be compared is valuable because an automatic search for optimised initial conditions or measurement points goes along with high computational costs. In any case improvements guided by scientific intuition can be checked quantitatively. This allows an effective planning of experimental studies used for parameter estimation before a sole experiment has been performed.

5 Conclusion

Almost every model in applied sciences like geodynamics, geomorphology, meteorology, soil science etc. needs empirical laws to describe natural systems. The parameters of those laws have to be fitted to measurements for meaningful results. We have seen that the reliability of such parameter estimations is not independent of the experimental design. If a model is used as basis for further modeling activities the consequences of wrong parameter estimation may be enormous.

Since the awareness of the problem does not seem to be widely spread, we have sought to find quantitative and generic methods to derive the information gain that a certain experiment can yield for the empirical parameters. Information theory has been presented as a general tool to determine optimised initial states and observation times for parameter estimation. It can be applied to all models that fit the form of general nonlinear dynamical systems (Eq. 16).

The methods have been demonstrated for a simple model that describes meander migration. It has been developed based on physical fluid dynamical laws (Navier–Stokes equation) and empirical parts that are only justified by visual inspection of the resulting evolution. Both parts contain empirical parameters, that can be estimated from observed evolutions of a natural system. It has been shown how the dimension of the model's state space can be reduced to obtain a complexity treatable for information methods.

Since the awareness about chaotic behaviour of dynamical systems rises, the requirements for a common acceptance of these methods is given. Especially in all geosciences, it should be common use to ask about the information a certain observation yields for estimation purposes because it is seldom possible to repeat experiments many times and typically the expenditure for a measurement is high.

Acknowledgment. This work has been kindly supported by the German Research Foundation (DFG) within the the Collaborative Research Center 350.

References

Ackers, P., and F.G. Charlton 1970): Dimensional analysis of alluvial channels with special references to meander length. Journal of Hydraulic Research, 8(3):287–316.

Allen, J.R.L (1992): Principles of physical sedimentology. Chapman & Hall.

Bernard, R.S., and M.L. Schneider (1992): Depth–averaged numerical modeling for curved channels. Technical Report HL-92-9, Dept. of Army, Waterways Experiment Station, Corps of Engineers, 3909 Halls Ferry Road, Vicksburg, Mississippi 39180–6199.

Carson, M.A., and M.F. Lapointe (1983): The inherent asymmetry of river meander planform. Journal of Geology, 91:41–55.

Cherry, D.S., P.R. Wilcock, and M.G. Wolman (1996): Evaluation of methods for forecasting planform change and bankline migration in flood–control channels. Technical report, Department of Geography and Environmental Engeineering, John Hopkins University, Baltimore, MD 21218.

Ferguson, R.I. (1975): Meander irregularity and wavelength estimation. Journal of Hydrology, 26:315–333.

Hjulström, F. (1942): Studien über das Mäander–Problem. Geografiska Annaler, 24:233–269.

Howard, A.D. (1992): Modeling channel migration and floodplain sedimentation in meandering streams. In: Carling, P.A., and G.E. Petts (eds.): Lowland floodplain rivers: geomorphological perspectives. Wiley & Sons Ltd.

Howard, A.D. (1995): Modelling channel evolution and floodplain morphology. private communication, to be published in 'Floodplain processes', edited by M. Anderson & D. Walling, Department of Environmental Sciences, University of Virginia, Charlottesville, VA 22903.

Hsü, K.J. (1989): Physical Priciples of Sedimentology. A readable textbook for beginners and experts, Springer Verlag.

Ikeda, S., G. Parker, and K. Sawai (1981): Bend theory of river meanders, 1. linear development. Journal of Fluid Mechanics, 112:363–377.

Johannesson, H., and G. Parker (1989): Linear theory of river meanders. In: Ikeda, S., and G. Parker (eds.): River meandering, American Geophysical Union.

Langbein, W.B., and L.B. Leopold (1966): River meanders – theory of minimum variance. Geological Survey Professional Paper 422–H, United States Government Printing Office, Washington.

Leopold, L.B., and W.B. Langbein (1966): River meanders. Scientific American, 214:60–70.

Leopold, L.B., and T. Maddock (1953): The hydraulic geometry of stream channels and some physiographic implications. Geological Survey Professional Paper, 252:1–57.

Malcherek, A. (1995): Numerische Methoden für Strömungen, Stoff– und Wärmetransport. http://www.hamburg.baw.de/hnm/nummeth/index.htm

Markham, A.J., and C.R. Thorne (1992): Geomorphology of gravel–bed river bends. In: Billi, P., R.D. Hey, C.R. Thorne, and R. Tacconi (eds.): Dynamics of gravel–bed rivers. John Wiley & Sons.

Papoulis, A. (1984): Probability, random variables and stochastic processes. Electrical Enginieering Series. McGraw–Hill Book Company, 2nd ed.

Parker, G., K. Sawai, and S. Ikedai (1982): Bend theory of river meanders, 2. nonlinear deformation of finite–amplitude bends. Journal of Fluid Mechanics, 115:303–314.

Parker, G. (1976): On the cause and characteristic scales of meandering and braiding in rivers. Journal of Fluid Mechanics, 76(3):457–480.

Schumm, S.A. (1963): Sinuosity of alluvial rivers on the Great Plains. Geological Society of America Bulletin, 74:1089–1100.

Shannon, C.E., and W. Weaver (1949): The mathematical theory of communication. University of Illinois Press.

Vosselman, G. (1992): Relational matching. Lecture Notes in Comp. Science, 628, Springer, Heidelberg.

Yang, C.T. (1971): On river meanders. Journal of Hydrology, 13:231–253.

Zimmermann, C. (1974): Sohlausbildung, Reibungsfaktoren und Sedimenttransport in gleichförmig gekrümmten und geraden Gerinnen. Dissertation, Fakultät für Bauingenieur- und Vermessungswesen der Universität Karlsruhe.

Subject Index

Springer
and the
environment

At Springer we firmly believe that an international science publisher has a special obligation to the environment, and our corporate policies consistently reflect this conviction.
We also expect our business partners – paper mills, printers, packaging manufacturers, etc. – to commit themselves to using materials and production processes that do not harm the environment. The paper in this book is made from low- or no-chlorine pulp and is acid free, in conformance with international standards for paper permanency.

 Springer

Lecture Notes in Earth Sciences

Vol. 57: E. Lallier-Vergès, N.-P. Tribovillard, P. Bertrand (Eds.), Organic Matter Accumulation. VIII, 187 pages. 1995.

Vol. 58: G. Sarwar, G. M. Friedman, Post-Devonian Sediment Cover over New York State. VIII, 113 pages. 1995.

Vol. 59: A. C. Kibblewhite, C. Y. Wu, Wave Interactions As a Seismo-acoustic Source. XIX, 313 pages. 1996.

Vol. 60: A. Kleusberg, P. J. G. Teunissen (Eds.), GPS for Geodesy. VII, 407 pages. 1996.

Vol. 61: M. Breunig, Integration of Spatial Information for Geo-Information Systems. XI, 171 pages. 1996.

Vol. 62: H. V. Lyatsky, Continental-Crust Structures on the Continental Margin of Western North America. XIX, 352 pages. 1996.

Vol. 63: B. H. Jacobsen, K. Mosegaard, P. Sibani (Eds.), Inverse Methods. XVI, 341 pages, 1996.

Vol. 64: A. Armanini, M. Michiue (Eds.), Recent Developments on Debris Flows. X, 226 pages. 1997.

Vol. 65: F. Sansò, R. Rummel (Eds.), Geodetic Boundary Value Problems in View of the One Centimeter Geoid. XIX, 592 pages. 1997.

Vol. 66: H. Wilhelm, W. Zürn, H.-G. Wenzel (Eds.), Tidal Phenomena. VII, 398 pages. 1997.

Vol. 67: S. L. Webb, Silicate Melts. VIII. 74 pages. 1997.

Vol. 68: P. Stille, G. Shields, Radiogenetic Isotope Geochemistry of Sedimentary and Aquatic Systems. XI, 217 pages. 1997.

Vol. 69: S. P. Singal (Ed.), Acoustic Remote Sensing Applications. XIII, 585 pages. 1997.

Vol. 70: R. H. Charlier, C. P. De Meyer, Coastal Erosion – Response and Management. XVI, 343 pages. 1998.

Vol. 71: T. M. Will, Phase Equilibria in Metamorphic Rocks. XIV, 315 pages. 1998.

Vol. 72: J. C. Wasserman, E. V. Silva-Filho, R. Villas-Boas (Eds.), Environmental Geochemistry in the Tropics. XIV, 305 pages. 1998.

Vol. 73: Z. Martinec, Boundary-Value Problems for Gravimetric Determination of a Precise Geoid. XII, 223 pages. 1998.

Vol. 74: M. Beniston, J. L. Innes (Eds.), The Impacts of Climate Variability on Forests. XIV, 329 pages. 1998.

Vol. 75: H. Westphal, Carbonate Platform Slopes – A Record of Changing Conditions. XI, 197 pages. 1998.

Vol. 76: J. Trappe, Phanerozoic Phosphorite Depositional Systems. XII, 316 pages. 1998.

Vol. 77: C. Goltz, Fractal and Chaotic Properties of Earthquakes. XIII, 178 pages. 1998.

Vol. 78: S. Hergarten, H. J. Neugebauer (Eds.), Process Modelling and Landform Evolution. X, 305 pages. 1999.

Vol. 79: G. H. Dutton, A Hierarchical Coordinate System for Geoprocessing and Cartography. XVIII, 231 pages. 1999.

Vol. 80: S. A. Shapiro, P. Hubral, Elastic Waves in Random Media. XIV, 191 pages. 1999.

Vol. 81: Y. Song, G. Müller, Sediment-Water Interactions in Anoxic Freshwater Sediments. VI, 111 pages. 1999.

Vol. 82: T. M. Løseth, Submarine Massflow Sedimentation. IX, 156 pages. 1999.